Study Guide for the Telecourse

Project Universe

Second Edition

by

Joel M. Levine

and

Richard T. Searles

for

Coast Community College District
Costa Mesa, California

Saunders College Publishing
Philadelphia Fort Worth Chicago San Francisco
Montreal Toronto London Sydney Tokyo

Coast Community College District

David A. Brownell, Chancellor
Coast Community College District

William M. Vega, President
Coastline Community College

Leslie N. Purdy, Director
Office of Alternative Learning Systems

David P. Stone, Assistant Dean

Michael V. Olds, Instructional Designer

Michael S. Werthman, Publications Editor

Judith Lindow McDuff, Assistant Publications Editor

Sandee Harden, Vice-President, Program Production
KOCE-TV, Channel 50

William G. Neill, Series Producer

Southern California Consortium

Sally V. Beaty, Director
Southern California Consortium

Peter F. Buffa, Associate Producer and Writer

Senior academic consultant and text author: George O. Abell, Professor of Astronomy, University of California, Los Angeles

Content advisor: David A. Pierce, Professor of Astronomy, El Camino College

Content advisor: Stephen P. Lattanzio, Assistant Professor of Astronomy and Planetarium Director, Orange Coast College

Cover photo: National Optical Astronomy Observatories

Project Universe, a telecourse, is produced by the Coast Community College District, KOCE-TV, Channel 50, and the Southern California Consortium.

ISBN 0-03-022714-3

Published by Saunders College Publishing

Printed in the United States of America
Published simultaneously in Canada
12 039 987654

Preface

To the Student

Welcome to the telecourse **Project Universe,** an introductory course in astronomy. Whether you are planning a career in astronomy, taking the course as part of an academic course for college credit, or taking it just for pleasure, we believe that you will find **Project Universe** interesting, entertaining, enriching, and inspiring.

Project Universe is designed according to the philosophy that astronomy, more than any other contemporary scientific discipline, seeks to understand the "grand plan" of nature — not just the vast expanse of the universe beyond Earth, but how Earth and its varied inhabitants fit into the cosmic scheme. Astronomy gives us a breathtaking view of nature, from particles smaller than the atom to giant superclusters of galaxies. As an introductory astronomy course, **Project Universe** goes beyond describing the contents of outer space. It examines the scientific method and considers both its promises and limitations. It answers many of our questions and poses new ones time after time, so that we are continually probing into the innermost secrets of time and space.

As with most classroom courses, **Project Universe** has a textbook, a study guide, assignments, and tests. The text is *Realm of the Universe,* fourth edition, by George O. Abell, David Morrison, and Sidney C. Wolff (Philadelphia: Saunders College Publishing, 1988). This book, *Study Guide for the Telecourse Project Universe,* will guide your study through the course, providing reading and viewing assignments, study activities, and practice test questions.

In addition, **Project Universe** has a special learning element that most classrooms do not have: a companion half-hour television program for each of the thirty lessons in the course.

Course Goals

The designer, academic advisors, and producer of this television course have specified fourteen major goals for students taking **Project Universe.** By the end of the course, students should be able to:

- State the basic properties of electromagnetic radiation.

- Review the history of astronomy associated with the study of planetary motion.

- Describe the basic characteristics of the sun, planets, moons, and minor members (comets, meteoroids, etc.) in the solar system.

- List and describe the types of information gathered by planetary probes.

- Explain the general theories of the origin and evolution of the solar system, including the theories of the origin of the moon.

- Describe the basic methods involved in determining the distances, masses, and sizes of stars.

- Describe the methods used by astronomers to measure and classify the brightness of stars.

- State the characteristics of gaseous nebulas, star clusters, and variable stars.

- Explain the theories involved in stellar energy generation and stellar evolution.

- Summarize the present scientific thoughts concerning the existence of extraterrestrial life.

- State the basic characteristics of our Galaxy and compare these with the basic characteristics of other galaxies.

- Explain the current theories of the origin of the universe and the observational tests for these theories.

- Explain the basic principles of the theory of relativity and how they apply to cosmology.

- Name and construct the hierarchy of objects in the observable universe.

How to Take This Television Course

If you are new to college courses in general, and television courses in particular, you can perhaps profit from a few suggestions offered by students who have successfully completed other television courses.

Television courses are designed for busy people — people with full-time jobs or family obligations — who want to take a course at home, fitting the study into their personal schedules. To do this, you will need to plan in advance how to schedule your viewing, reading, and study. Buy the books before the course begins and look them over; familiarize yourself with any materials supplied by your college and estimate how much time it will take you to complete special tests and assignments for each lesson. Write the dates of midterms, finals, review sessions, and special projects in your calendar so that you can plan to have extra time to prepare for them. Discuss with your family and friends your needs for study and television viewing time. Although you may find it enjoyable and instructive to watch the programs with other people, save the talking and discussion until *after* the program so that you won't miss important information. You may find it helpful to take five minutes at the end of the program to write a brief summary of what you have seen, the meaning of key concepts and terms, and any questions you may have about the program topic.

This study guide will be especially helpful to you as you progress through the course. Begin your study of each lesson by reading the lesson Overview, the Learning Objectives, and the Assignments for the lesson. Follow the sequence of reading and viewing recommended there and then use the Study Activities section as a way to review and study the information in the lesson. Note: Answers to certain questions in the "Using What You've Learned" section are in the answer key in Appendix 1. As a final step, take the Self-Test and check your answers with the answer key at the end of the lesson.

To further ensure that you will complete **Project Universe** to your satisfaction, you should heed these additional suggestions:

- Buy your own copies of the text and study guide for this course. You will not be able to pass the course without following the assigned exercises in these books.

- Watch all of the television programs. You will need to study both the written and television lessons to complete the tests and assignments.

- Keep up with your work *every week.* Set aside time for reading, viewing, and study; then stick to your schedule.

- Contact the instructor at your college or university if you have questions, need advice or help, or wish to review television programs or watch ones you've missed. More than likely, broadcasts will be repeated at least once by your local station, and videocassette copies of programs are generally available through a media center or library.

About the Television Program

Project Universe consists of thirty half-hour, full-color television programs especially designed and produced as a college-level course by the Coast

Community College District, with KOCE-TV (Channel 50). and the Southern California Consortium. Originally produced in 1978, the television component and all print materials for **Project Universe** have been periodically revised to conform with the most recent information and research in astronomy. The series makes extensive use of active astronomers to demonstrate and explain their fields of intererst and their research, with Dr. Edwin Krupp of the Griffith Observatory in Los Angeles serving as program host.

The telecourse is a basic, yet comprehensive, introduction to the evolving science of astronomy. **Project Universe** uses the unique capabilities of television to introduce students to the origin, characteristics, and evolution of the solar system, the stars, the galaxies, and the universe. Among the varied techniques employed in the television programs are animation, special effects, technical films, and interviews with some of the most noted research scientists in astronomy and related fields. Some of the topics include: the history of astronomy; the nature of starlight; characteristics and features of all the planets; comets and meteoroids; the birth and death of stars; black holes; the nature of the Milky Way; quasars; and cosmology.

We hope that you will enjoy this course and will come to a new and deeper understanding and appreciation of astronomy.

Office of Alternative Learning Systems
Coastline Community College

Acknowledgments

Producing the **Project Universe** television course has been a complex team effort by many people with many skills. Several of those persons respoonsible for this course are listed on the copyright page of this book.

We are deeply grateful to the many members of the astronomical community who have generously shared their time and expertise, and to the many institutions that have shared their resources. We wish to give special acknowledgment to NASA and its Jet Propulsion Laboratory, to the California Institute of Technology, to Griffith Observatory for its strong commitment to public education, and to the late George O. Abell for many of the activities contained in his *Instructor's Manual for George O. Abell, Realm of the Universe,* second edition, prepared by James Becker (Philadelphia: Saunders College/Holt, Rinehart and Winston, 1980). Programs for the revised telecourse were produced in the studios of the Southern California Consortium, Hollywood, California. This study guide, test materials, and instructor's materials for the course were developed for publication by the Office of Alternative Learning Systems, Coastline Community College, Fountain Valley, California, one of the three colleges in the Coast Community College District.

Contents

The Astronomer's Universe

1

Overview

In June 1983, a rather small machine drifted past the orbit of the outermost planet in our solar system. Pioneer 10 had been launched from Earth in 1972 — it took eleven years to reach the edge of the solar system. And, as you'll learn in your viewing and reading for this lesson, its journey is just beginning. In a small and symbolic way, however, humankind might be said to have begun participating in the larger universe, rather than simply observing it.

Astronomy is a science which attempts to understand and describe the materials and processes found in the universe. It deals with matter, emptiness, and time on both the smallest and grandest scales, and with distances of such magnitude that astronomers measure them in units of time rather than length. Even those units of time are mind-boggling. Some of the light and other radiation observed by astronomers today has travelled so far that we literally see almost back to the beginning of time itself.

Interest in the heavens began long before history was first written. Human beings began to study the heavens, trying to fathom the secrets of the cosmos, even before Greek civilization reached its zenith. Some of the discoveries and deductions by Greek thinkers have been proven accurate, although many proved to be erroneous as new evidence was studied centuries later during the Renaissance. In fact, much of what we know about the universe, and even about our own solar system, has been learned only in the twentieth century. So while taking the course, you are not only participating in one of our oldest sciences, you're also learning about some of humankind's newest discoveries and technologies.

In this lesson, you will receive a whirlwind introduction to the field of astronomy and the work of its scientists. Both your reading assignments and the television program take you on a brief "grand tour" of the universe as it is perceived by science today. Incidentally, it will be pointed out to you that modern astronomers are not searching for absolute "truth," but rather are trying to construct accurate "models" of the universe which are consistent with all the observable evidence. The current model could be (and probably will be) modified by new evidence and new theories in the future. Keep in mind that most of the universe you'll see in your grand tour has been discovered only in this century.

In the program for this lesson you'll be shown some demonstrations of scientific principles that apply throughout the universe and which are employed by astronomers in the study of stars and galaxies and other structures that populate space.

Learning Objectives

When you have completed all assignments in this lesson, you should be able to:

1. Beginning with our solar system, name the major structures in the universe and list them in their order of size and distance from Earth.

2. Describe and explain what the scientific method is and how it is used in the science of astronomy.

3. Describe the basic motions of the planets and the various motions of Earth in particular.

4. Describe the technique Eratosthenes used to measure the circumference of Earth.

Assignments for This Lesson

Before Viewing the Program

Read the overview at the beginning of this lesson and look over the learning objectives that precede the assignments.

Read the Prologue and Chapter 1 in the text.

Skim the text section "Astrology," pages 23-26.

View Program 1, "The Astronomer's Universe"

As you view the program, look for:

- the identification and description of the different structures in the universe.

- the discussion of and attempt to visually demonstrate the distance scale of the universe.

- the discussion of how the laws of science that were discovered on Earth can be used to explain distant astronomical phenomena.

After Viewing the Program

Review the text assignments.

Review the terms that appear in italics or boldfaced type in the text. In particular, note the key terms and concepts listed in the study activities section.

Complete the study activities and the self-test in this study guide; then check your answers with the answer key at the end of the lesson.

Complete the following exercises:

1. Make a list of the major structures in the universe and place them in order according to size, starting with the smallest (the atom) and ending with the largest (cluster of galaxies).

2. Two theories compete to explain a particular phenomenon. One represents experimental data well; the other is more consistent with our

ideas on what the universe is like. Which theory is better? Write down the reasons in support of your answers.

3. Complete exercises 4 and 6 in text Chapter 1, page 30.

Study Activities

Key Terms and Concepts

This lesson and all lessons that follow contain vocabulary terms and concepts with which the student should become familiar. Terms and concepts are defined in the glossary from the text for **Project Universe** and/or the glossary in the back of this study guide.

solar system
astronomical unit
light-year
Milky Way
atomic nucleus
parallax
celestial sphere
celestial pole
constellation
epicycle
deferent
eccentric
heliocentric
geocentric

Using What You've Learned

Answers to the questions in this section for this lesson and others can be found in the back of this study guide.

1. We have become a very sheltered civilization. We no longer look at the location on the horizon where the sun rises to determine when to plant the grain. Nor do we watch the changes in the phases of the moon to determine the beginning of a new month. Today we use books like the *Old Farmer's Almanac* to tell us when to plant, or a calendar to tell us what month it is. We have lost touch with nature. Your assignment is to go outside and re-familiarize yourself with the sky.

   a. Sketch the eastern or western horizon to the best of your ability. Record on your drawing the point where the sun rises (east) or sets (west) once a week for the duration of the semester. What do you observe?

   b. Identify three or four constellations visible to you at about 9:00 P.M. Do the same at the end of the semester. What do you notice? Why?

c. Locate the planets visible in the sky over the course of an entire evening. Select one and follow its motion against the background stars for a period of several weeks.

d. Observe the moon once or twice a week for a period of one month. Sketch the appearance of the moon as you see it.

2. With a pair of binoculars, locate (if possible) the following objects and sketch their appearance with and without the use of binoculars: the moon, a star, the Milky Way, Venus, Jupiter, Saturn, and Mars. DO NOT OBSERVE THE SUN.

3. Visit a planetarium. Planetariums and observatories offer visual and entertaining supplements to your study of astronomy. Take a look at the exhibits that are on display. If possible, look through the telescope. Your visit to a planetarium will introduce you to many of the topics you will be studying in this course.

Recommended Readings

For supplementary reading see the bibliography in the back of this study guide.

Self-Test

(Select the one best answer.)

Objective 1: Beginning with our solar system, name the major structures in the universe and list them in their order of size and distance from Earth.

1. Telescopes have reached out to distances of 10 billion light-years. List, in order of decreasing distance, the following objects beginning with the farthest.
   a. Andromeda galaxy
   b. the moon
   c. Alpha Centauri
   d. the sun
   e. distant cluster of galaxies
   f. the planet Neptune
   g. artificial satellite in orbit around Earth

2. Of the following, which phenomena can be easily observed with the naked eye?
   a. constellations seen by the Greeks of 500 B.C.
   b. changes in the positions of the stars in the ancient constellations of 500 B.C.
   c. changing phases of the moon
   d. motion of the planets against the background stars

Objective 2: Describe and explain what the scientific method is and how it is used in the science of astronomy.

3. From the following list, select the advances in astronomy that have so far been made.
   a. the discovery of extraterrestrial life
   b. the use of telescopes for photography
   c. the use of nonvisible radiation to study objects
   d. the exploration of the moon and planets by spacecraft

4. Which of the following is (are) required by the scientific method?
   a. logical argument
   b. physical or observational evidence
   c. consistency
   d. testability
   e. all of the above

Objective 3: Describe the basic motions of the planets and the various motions of Earth in particular.

5. For a short period of time a planet will be observed to move westward instead of eastward against the fixed stars. This temporary motion is called
   a. precession.
   b. parallax.
   c. retrograde motion.
   d. annual motion.
   e. daily motion.

6. The fact that early scientists could not detect stellar parallax led them to believe that the
   a. Earth orbited the sun.
   b. moon orbited Earth.
   c. Earth was motionless in space.
   d. planets moved against the background stars.

7. Ancient people observed all of the following except
   a. the planet Saturn.
   b. the phases of the planet Venus.
   c. the phases of the moon.
   d. the satellites of Jupiter.
   e. the rising of the sun, moon, and stars.
   f. meteors and comets.

8. Planets, as seen by the naked eye,
   a. are located near the ecliptic.
   b. twinkle less than stars.
   c. are generally brighter than stars.
   d. move with respect to the apparently stationary stars.
   e. are described by all of the above characteristics.

Objective 4: Describe the technique Eratosthenes used to measure the circumference of Earth.

9. The unit of measure Eratosthenes used to measure the circumference of Earth was the
   a. kilometer.
   b. mile.
   c. foot.
   d. yard.
   e. stadium.

10. Eratosthenes measured Earth's size by
    a. noting the times of the first-quarter and last-quarter moons.
    b. studying the times and durations of solar and lunar eclipses.
    c. measuring the different zenith angle of the sun at noon on the first day of summer as seen from two different places on Earth.
    d. assuming that Earth moves in an ellipse around the sun.
    e. assuming that the sun moves around Earth.

Answer Key

1. e, a, c, f, d, b, g
2. a, c, d
3. b, c, d
4. e
5. c
6. c
7. b, d
8. e
9. e
10. c

Historical Perspectives 2

Overview

Humankind has probably always enjoyed the sense of wonder that accompanies the viewing of stars on a dark night. As you learned in your reading for the preceding lesson, even the oldest civilizations felt the need to explain the stars and Earth's place within the heavens. It isn't hard to understand why the ancients placed Earth at the center of everything, is it? Even today, we speak of the sun's "rising" and "setting"; we follow the "path" of the sun or stars "across" the sky; and we can locate stars by the patterns we call constellations, as if these stars were actually grouped together. When you look at the clear night sky, don't the stars and planets look as if they were really embedded together in a vast inverted bowl enclosing the earth? If all you had was the evidence of your senses, could you believe other than that Earth is flat and the heavens pass overhead in one or more spheres?

This lesson completes the history, begun in the first chapter of your text, of how thinkers and scientists came to understand the true organization of the sun and its planets. In your reading and in the television program you'll learn about the contributions of some of the greatest minds in history: Copernicus, who resurrected the idea of the sun being at the center of the universe; Tycho Brahe, whose skills in precise observation provided new and important data about the movement of planets; Galileo, who was the first to make astronomical observations with a telescope; Johannes Kepler, who formulated three laws of planetary motion to describe the orbits of bodies around the sun; and Isaac Newton, who expanded upon Kepler's laws and described the properties of matter, such as mass, acceleration, and gravitation, that govern the movements of objects in orbit.

Some of the material in this lesson may require extra study. The concepts underlying Kepler's laws and Newton's laws and the theory of gravitation are actually quite simple, but they are not phenomena that we encounter every day. If these concepts are new to you, take the time necessary to clearly understand each concept, and work to add such terms as momentum, acceleration due to gravity, and ellipse to your working vocabulary. The explanations in the text and animated art in the television program will also help to clarify these concepts. Throughout this course you will encounter aspects of these laws that relate to motion.

This lesson also discusses some of the effects of Earth's movements in its journey around the sun. For example, Earth's rotation makes the stars and planets appear to be changing position continuously, so that a system is needed to locate their exact positions in the sky. You will be introduced to one system used by astronomers to identify such positions. You'll also learn that astronomers refer to a "day" which is slightly shorter than our familiar "sun time" solar day.

Every schoolchild today learns something about the rotation of Earth, the organization of the solar system, and the phenomenon of gravity. These concepts now seem natural because we become familiar with them at an early age. But civilized and educated people not many centuries ago felt that failure to believe the universe revolved around Earth was a matter of life and death. In fact, Galileo saved his own life only by recanting his belief in a heliocentric system. Educated people today must know about Galileo and such men as Copernicus, Kepler, and Newton, who studied the planets and stars and derived new knowledge from what they found.

Learning Objectives

When you have completed all assignments in this lesson, you should be able to:

1. Describe the heliocentric model of the universe as proposed by Copernicus and perfected by Newton and compare it to the geocentric model of the universe as developed by the classical Greek philosophers and astronomers.
2. State and explain Kepler's three laws of planetary motion
3. Describe the important telescopic observations of Galileo.
4. State Newton's laws of motion and law of universal gravitation and describe how they explain orbital motion.
5. List and describe the various coordinates both on Earth and in the sky — right ascension and declination, azimuth and altitude — and describe how these are used to orient an earthbound observer to celestial phenomena.
6. Describe and explain the differences between sidereal, solar, and standard time.
7. Describe the various motions Earth undergoes and the observed effects produced by these motions.

Assignments for This Lesson

Before Viewing the Program

Read the overview at the beginning of this lesson and look over the learning objectives that precede the assignments.

Read Chapter 2 in the text, pages 32-49.

Reread text Chapter 1, pages 26-30.

Read carefully the following sections in text Chapter 3, pages 50-60: "Earth and Sky," "The Seasons," "Time," and "Date."

View Program 2, "Historical Perspectives"

As you view the program, look for:

- the solar system models of Aristarchus, Ptolemy, and Copernicus.
- the observations of Galileo and how these observations changed our understanding of the solar system and the universe.
- a detailed explanation of Kepler's three laws.
- the explanation and significance of Newton's three laws of universal gravitation.
- the explanation and significance of Newton's three laws of motion.

After Viewing the Program

Review the text assignments.

Review the terms that appear in italics or boldfaced type in the text. In particular, note the key terms and concepts listed in the study activities section.

Complete the study activities and the self-test in this study guide; then check your answers with the answer key at the end of the lesson.

Complete the following exercises: 4, 7, 8, and 9 in text Chapter 2, pages 48-49; and 5 and 8 in text Chapter 3, page 68.

Study Activities

Key Terms and Concepts

meridian
hour circles
zenith
right ascension
declination
altitude
azimuth
equinox
vernal equinox
solstice
sidereal time
mean sun
mean solar time
apparent solar time
universal time
sidereal month
synodic month
Julian calendar
Foucault pendulum
weight
mass
ellipse
universal gravitation
inertia

Using What You've Learned

1. In the following diagram of a planetary orbit locate:
   a. one possible position of the sun
   b. the major axis
   c. the position where the planet would have its slowest speed
   d. the perihelion

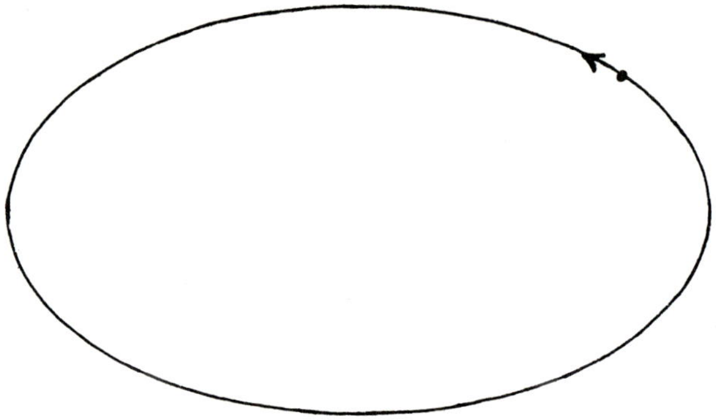

Figure 2.1. Planetary orbit.

2. An astronaut in space trying to repair a satellite gets frustrated and throws a wrench away. Assume that he or she isn't holding on tightly to the satellite or anything else. What happens to the astronaut? Would you expect him or her to move slower or faster than the wrench?

3. Where on Earth:
   a. do the stars rise in the west?
   b. is the south celestial pole at the south point on the horizon?
   c. is the sun at the north point on the horizon at midnight on June 22?
   d. is the zenith directly overhead?
   e. does the summer solstice pass through the zenith on June 22?
   f. is the observer's latitude equal to the altitude of the north celestial pole (assume the observer is in the northern hemisphere)?
   g. does the sun rise once a year, on Sept. 23?
   h. are all the stars above the horizon half the time?

Recommended Readings

For supplementary reading see the bibliography in the back of this study guide.

Self-Test

(Select the one best answer.)

Objective 1: Describe the heliocentric model of the universe as proposed by Copernicus and perfected by Newton and compare it to the geocentric model of the universe as developed by the classical Greek philosophers and astronomers.

1. In the solar system according to ______________, the sun is located beyond the orbs of Mercury and Venus and moves around Earth once a day.
   a. Galileo
   b. Aristarchus
   c. Ptolemy and Hipparchus
   d. Copernicus

2. The rotation of Earth on its axis and the revolution of Earth around the sun were central to the modern theory of planetary motions first made by
   a. Copernicus.
   b. Kepler.
   c. Galileo.
   d. Newton.
   e. Ptolemy.

Objective 2: State and explain Kepler's three laws of planetary motion.

3. According to Kepler, the orbits of the planets are ______________, with the sun at ______________.
   a. ellipses, one of the foci
   b. ellipses, the center
   c. circles, the center
   d. circles, the deferent

Objective 3: Describe the important telescopic observations of Galileo.

4. Which of the following observations were made by Galileo?
   a. craters and mountains on the moon
   b. phases of Venus
   c. moons around Jupiter
   d. stars within the Milky Way
   e. all of the above

Objective 4: State Newton's laws of motion and law of universal gravitation and describe how they explain orbital motion.

5. Which of the following correctly expresses Newton's law of universal gravitation?
   a. $F = G \frac{m_1 m_2}{d}$
   b. $F = G \frac{(m_1 + m_2)}{d^2}$
   c. $F = G \frac{m_1 m_2}{d^2}$
   d. $F = ma$
   e. $F = G \frac{(m_1 - m_2)}{d^2}$

6. If the mass of one of two objects is tripled and the distance between the two objects is also tripled, the force of gravity is
   a. 1/9 as great
   b. 1/3 as great
   c. 3 times greater
   d. 9 times greater
   e. 27 times greater

7. The weight of any object is the
   a. amount of space it takes up.
   b. quantity of matter in the object.
   c. force of gravity upon the object.
   d. same as its density.

Objective 5: List and describe the various coordinates both on Earth and in the sky — right ascension and declination, azimuth and altitude — and describe how these are used to orient an earthbound observer to celestial phenomena.

8. The position of a star on the celestial sphere measured above or below the celestial equator is called the
   a. right ascension.
   b. azimuth.
   c. altitude.
   d. declination.

Objective 6: Describe and explain the differences between sidereal, solar, and standard time.

9. The sidereal day is defined in terms of the
   a. rotation of Earth with respect to the sun.
   b. rotation of Earth with respect to the stars.
   c. motion of the sun along the ecliptic.
   d. motion of the vernal equinox along the celestial equator.

Objective 7: Describe the various motions Earth undergoes and the observed effects produced by these motions.

10. The observation that Venus went through phases and that Jupiter had satellites in orbit around it eventually led scientists to conclude that the
    a. sun orbited Earth.
    b. Earth orbited the sun.
    c. solar system contained six planets.
    d. solar system is best described by a geocentric theory.

11. If you were standing at the north pole of Earth you would observe
    a. the stars rising in the east and setting in the west.
    b. the stars rising in the west and setting in the east.
    c. the stars appearing to make circular paths around the star Polaris.
    d. no stars at all.

Answer Key

1. c	7. c
2. a	8. d
3. a	9. b
4. e	10. b
5. c	11. c
6. b	

Lunar Aspects 3

Overview

For centuries it was common for people to consult their celestial calendars — or the local astrologer or astronomer — for guidance about almost everything they did and thought. Most people of the industrial and post-industrial age are only casual observers of the heavens. Few of us are able to "read the sky" as our agricultural ancestors did. Instead, we mark the passage of time by reading our digital watches, newspapers, and television schedules.

There are exceptions. The moon, Earth's nearest celestial neighbor, continues to fascinate us. Although visits to the moon by Apollo mission astronauts may have diminished its romantic aura, Americans have certainly acquired a greater scientific familiarity with and understanding of this heavenly body. "Moon rock" exhibits continue to attract curious crowds, the moon's surface features are frequent subjects of posters and illustrations, and many of us have witnessed the lesser weight of objects on the moon's surface from television pictures taken during the "moon walks."

Most of us don't bother to follow the daily progression of the moon through its phases. We still admire an occasional full moon on a particularly clear night, but city lights allow "new" moons to come and go unnoticed. Lunar eclipses usually receive brief mention in local news programs, but most people don't bother to watch such phenomena in their entirety, if at all.

A total solar eclipse continues to be one of nature's most awesome events, however. Even today, when the fear and mystery of the event have been dispelled, people — and not just astronomers — will travel thousands of miles at great expense to observe an occurence that lasts a little over seven minutes at best, and that may be obscured altogether by clouds. Astronomers are interested in solar eclipses because they permit us to view objects and events not otherwise observable: the tenuous outer atmosphere of the sun (the corona), normally overwhelmed by the brilliance of the solar disk; the bending of stellar light rays as they pass close to the sun; and objects that might be moving close to the sun itself. Other people watch a solar eclipse for the sheer grandeur and majesty of the phenomenon.

As you have already learned, the plane of Earth's orbit is also called the ecliptic plane. The orbit of the moon does not lie in the ecliptic plane. The inclination (tilt) of the moon's orbit relative to the ecliptic plane is about 5°. For this reason, the moon does not line up "exactly" with Earth and the sun.

As the moon orbits Earth, it lies above or below the ecliptic plane most of the time. Only twice a month does the moon cross the plane of Earth's orbit and it is only at or near these times that an eclipse can occur. A second requirement for an eclipse is that these crossings occur at or near new or full phase of the moon.

Only if the two conditions described in the above paragraph occur at or near the same time can a solar or lunar eclipse take place. Since these so-called eclipse seasons do not happen every month but are approximately six months apart, eclipses do not happen every month. In fact, a maximum of seven eclipses can occur in any given year.

This lesson deals with lunar dimensions and motions, the moon's appearance in the sky, and both solar and lunar eclipses. You will come to understand why the moon appears as it does in the sky and the nature of the moon's orbit. You will also learn about the moon's several motions; its movement about its own axis (rotation) and around Earth (revolution), and its movements with respect to the sun (synodic) and other stars (sidereal).

In the television program and the text, the mechanics of a total solar eclipse are carefully detailed, step by step, and the television program provides dramatic footage of an actual eclipse viewed from a desert in Mexico. You will come to understand how such fleeting events can be predicted with pinpoint accuracy years — even centuries — in advance. The differences between a total, a partial, and an annular eclipse are also explained. You will also study lunar eclipses and understand why, even though these are less common than their solar counterparts, they are more readily experienced by earthbound observers.

Figure 3.1 and Table 3.1 give the times and locations of visibility for solar and lunar eclipses during the next few years. The next total solar eclipse visible from the continental United States will not occur until well into the next century. However, a total solar eclipse path will cut across Hawaii, Mexico, and the intervening Pacific Ocean on July 11, 1991. This eclipse should be reasonably accessible to Americans, particularly for passengers on chartered cruise ships which can travel to the most likely sites for clear weather.

It may seem paradoxical that whereas eclipses of the moon are slightly less frequent than solar eclipses, you are much more likely to observe a lunar eclipse. This is because when the moon moves into Earth's shadow, roughly half the people on Earth can observe the event. But when the sun is eclipsed, one must be within a very narrow path (the solid lines on Figure 3.1) to observe the event in totality.

Lunar eclipses are of lesser scientific interest than solar eclipses. They are colorful events, however, and easy to observe, since no equipment is required and you can usually observe them from right outside where you live. Very interesting photographs of lunar eclipses can be obtained with little more than a hand-held camera. You should plan to take a series of photos of the next lunar eclipse.

Understanding the mechanics of eclipses in the sun-Earth-moon system allows astronomers to use similar phenomena — occultations and transits — to study celestial bodies in other parts of the solar system and in even deeper space. The lessons of lunar and terrestrial motions become the tools for an even greater understanding of the nature of the universe.

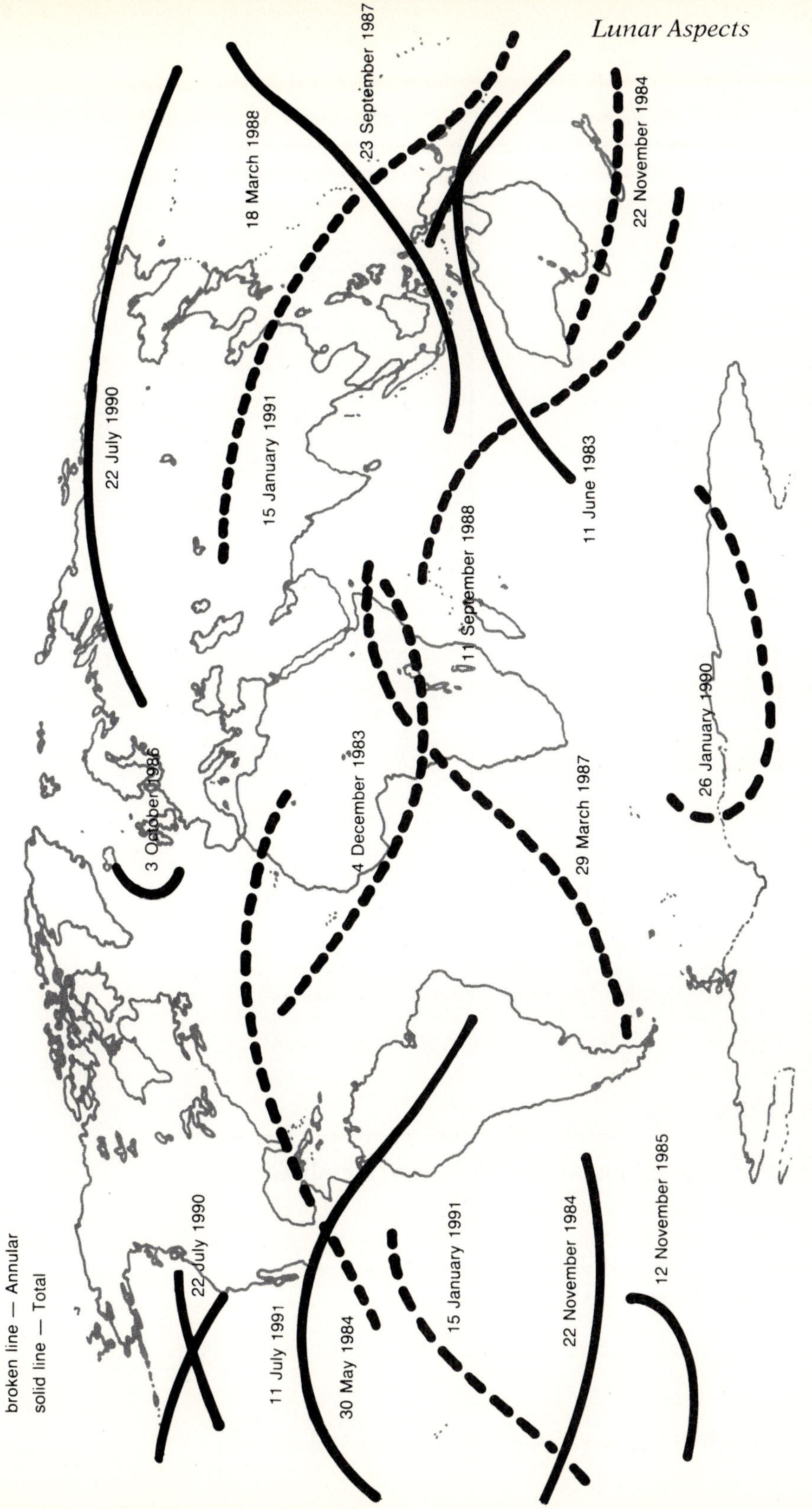

Figure 3.1. Total (____) and annular (-----) solar elipse paths for the period 1983 through 1991.

Date and Time	Total or Partial	Where Visible
1988, 3 March 11:07 A.M. 11:14 A.M. 11:21 A.M.	Partial	Extreme NW only.
1988, 27 Aug 5:08 A.M. 6:06 A.M. 7:30 A.M.	Partial	Beginning all over U.S. except NE, end in West only.
1989, 20 Feb 8:44 A.M. 10:36 A.M. 12:28 P.M.	Total	Begin W; end NW.
1989, 16 Aug 8:22 P.M. 10:09 P.M. 11:57 P.M.	Total	Begin E half of U.S., end all of U.S. except NW.
1990, 9 Feb 12:30 P.M. 2:12 P.M. 3:54 P.M.	Total	Beginning extreme NW only.
1990, 6 Aug 7:45 A.M. 9:13 A.M. 10:42 A.M.	Partial	Beginning extreme W only.
1991, 29/30 Jan 10:59 P.M. 1:00 A.M. 2:00 A.M.	Penumbral	All over U.S.
1991, 26 June 8:47 P.M. 10:16 P.M. 11:44 P.M.	Penumbral	Begin Eastern half, end all over U.S. except NW.
1991, 21 Dec 5:01 A.M. 5:34 A.M. 6:07 A.M.	Partial	All over U.S.

Table 3.1. Partial and total lunar exlipses visible in continental U.S., 1988-1991.

Dates and times are Eastern Standard. Recognize, however, that these times may differ slightly depending on the geographical location of the observer in the United States and the time zone he or she is in. Three times listed are: first contact with umbra, maximum eclipse, last contact with umbra.

Learning Objectives

When you have completed all assignments in this lesson, you should be able to:

1. Describe the methods for determining the distance to the moon, the diameter of the moon, and the mass of the moon.

2. Describe the lunar orbit and how it is oriented with respect to the ecliptic.

3. Define and explain the differences between the sidereal and synodic months.

4. Draw a diagram showing the cause of the various phases of the moon, and name these phases.

5. Describe the rotation of the moon.

6. Explain how the moon influences Earth's ocean tides, and describe the concept of tidal forces.

7. Describe and explain the circumstances resulting in the occurrence of various types of solar eclipses.

8. Cite a number of scientific reasons for observing total solar eclipses.

9. Describe and explain the circumstances resulting in the occurrence of various types of lunar eclipses.

Assignments for This Lesson

Before Viewing the Program

Read the overview at the beginning of this lesson and look over the learning objectives that precede the assignments.

Review Chapters 1 through 3 in the text, pages 15-67. In particular, reread the section on "Tides," pages 44-48, and pages 60-67.

View Program 3, "Lunar Aspects"

As you view the program, look for:

- the illustration of the phases of the moon.

- an explanation of the "man in the moon," "harvest moon," and "moon illusion."

- an explanation of the sidereal and synodic months.

- a description of the lunar laser ranging experiment.

- an explanation of the occurence of the different types of lunar and solar eclipses and why eclipses do not occur every month.

After Viewing the Program

Review the text assignments.

Review the terms that appear in italics or boldfaced type in the text. In particular, note the key terms and concepts that are listed in the study activities section.

Complete the study activities and the self-test in this study guide; then check your answers with the answer key at the end of the lesson.

Complete the following exercises: 13, 14, 15, and 18, on page 68 in text Chapter 3.

Study Activities

Key Terms and Concepts

phases of the moon
sidereal period
synodic period
nodes
total eclipse
partial eclipse
annular eclipse
Baily's beads
occultation
transit
eclipse seasons

Using What You've Learned

1. Observe the moon for as many nights during a phase cycle as possible. Make note of (a) the rising and setting times, (b) the motion of the moon against the background stars, and (c) the phase of the moon.

2. Photograph the moon each night for as much of the phase cycle as possible. (Record the dates.) Be sure to use the same lens for all exposures. (You will need only 1/125 exposure time at 400 ISO for most of these photographs. Check your meter to be sure.) Compare the diameter of the moon in each photograph and determine the dates of apogee and perigee. (The moon will be largest at perigee.)

3. Make a list of the sources that give information concerning the dates, times, and locations of lunar and solar eclipses that will occur during the next year.

4. Compare the differences between terrestrial and lunar environments. List the reasons for these differences.

5. Why can't the total phase of an eclipse of the sun last an hour? What is the maximum time for a solar eclipse?

Recommended Readings

For supplementary reading see the bibliography in the back of this study guide.

Self-Test

(Select the one best answer.)

Objective 1: Describe the methods for determining the distance to the moon, the diameter of the moon, and the mass of the moon.

1. The average Earth-moon distance is approximately ______________ kilometers.
   a. 384
   b. 300,000
   c. 384,400
   d. 93,000,000

2. The most recent estimate of the mass of the moon was made by observing
   a. satellites in orbit around the moon.
   b. phases of the moon.
   c. sunspots.
   d. moon rocks brought back by the Apollo astronauts.

3. Earth is ______________ times more massive than the moon.
   a. two
   b. four
   c. sixty-four
   d. eighty-one

4. The diameter (in kilometers) of the moon is approximately
   a. 2,000.
   b. 3,470.
   c. 12,756.
   d. 240,000.

Objective 2: Describe the lunar orbit and how it is oriented with respect to the ecliptic.

5. The fact that the lunar orbit is inclined with respect to the Earth-sun plane prevents
   a. a lunar or solar eclipse from occuring every two weeks.
   b. a lunar eclipse from occurring twice a month.
   c. a solar eclipse from occurring twice a month.
   d. nothing at all.

Objective 3: Define and explain the differences between the sidereal and synodic months.

6. One revolution of the moon relative to the stars is called the
   a. phases of the moon.
   b. synodic month.
   c. sidereal period.
   d. conjunction.

7. One revolution of the moon relative to the Earth-sun line is called the
   a. synodic period.
   b. phases of the moon.
   c. sidereal period.
   d. opposition.

Objective 4: Draw a diagram showing the cause of the various phases of the moon, and name these phases.

8. The new phase of the moon occurs when the moon is
   a. at right angles to the Earth-sun line.
   b. between the sun and Earth.
   c. behind Earth.
   d. behind the sun.

9. The moon would be at full phase if it
   a. set at midnight.
   b. rose at midnight.
   c. set at sunrise.
   d. rose at sunset.

10. If the moon as seen from Earth is full, Earth as seen from the moon would be
    a. full.
    b. new.
    c. in the first quarter.
    d. a crescent.

Objective 5: Describe the rotation of the moon.

11. Because of the moon's synchronous rotation we see
    a. all of the lunar surface from Earth.
    b. only about half of the lunar surface from Earth.
    c. half of the lunar surface during the first two weeks of the month and the other half during the last two weeks of the month.
    d. None of the above is correct.

Objective 6: Explain how the moon influences Earth's ocean tides, and describe the concept of tidal forces.

12. The sun, moon, and ________________ are primarily responsible for the ocean tides.
    a. Jupiter
    b. Venus
    c. Earth
    d. Mars

Objective 7: Describe and explain the circumstances resulting in the occurrence of various types of solar eclipses.

13. If the moon is new and near one of the points of its orbit where it crosses the plane of Earth's orbit (a node), a(n)
    a. partial lunar eclipse would occur.
    b. total lunar eclipse would occur.
    c. annular eclipse may occur.
    d. total solar eclipse may occur.
    e. partial solar eclipse may occur.

14. A ring of sunlight is seen around the disk of the moon during an eclipse of the sun. This is called a(n)
    a. partial solar eclipse.
    b. partial lunar eclipse.
    c. total solar eclipse.
    d. annular eclipse.

Objective 8: Cite a number of the scientific reasons for observing total solar eclipses.

15. As seen from Earth, what two objects have an angular size of ½°?
    a. the moon
    b. the sun
    c. Jupiter
    d. Venus

Objective 9: Describe and explain the circumstances resulting in the occurrence of various types of lunar eclipses.

16. The angular size of Earth is 2° as seen from
    a. the sun.
    b. Jupiter.
    c. the moon.
    d. Mars.

Answer Key

1. c	9. c, d
2. a	10. b
3. c	11. b
4. b	12. c
5. a	13. c, d, e
6. c	14. d
7. a	15. a, b
8. b	16. c

Electromagnetic Radiation

4

Overview

Modern scientists can show indisputable evidence that the universe is vast. The ancients sensed this without knowing just how huge the "celestial sphere" really is and without the use of sufficiently sensitive or precise measuring instruments. All they had was the sense of sight, and with this they perceived the light from the stars. Even today, "starlight" is the only way most of us will ever see the objects of our universe. Our other senses are useless in viewing faraway objects because of the extreme distances involved. Moreover, our sense of sight is imperfect and the light from the heavens extremely faint.

However faint, starlight is astronomy's main key to unlocking the secrets of the universe, and astronomer's use it wisely. By using telescopes, astronomers have collected and focused starlight. With other, increasingly sensitive, instruments they have measured starlight, amplified it, dispersed it, analyzed its constituent parts, and studied its properties and effects under a variety of conditions. Scientists have discovered that light is a form of radiant energy and that it is perceived by the human eye because the energy in light stimulates a sight response in certain eye nerves. What appears to be uniformly "white" light from the sun and stars is actually a mixture of colors. By passing this light through a prism, its constituent colors — collectively called the *spectrum* of light — can be examined separately and much of the behavior of light, and the chemical characteristics of the light's source, described and explained. For example, when sunlight passes through a prism, it is broken up into the colors of the rainbow. However, upon close examination of the solar spectrum, numerous dark lines can be seen crossing the continuous blend of colors. The German physicist, Joseph von Fraunhofer (1787-1826), mapped nearly 600 of the most prominent lines, now called Fraunhofer lines.

Fraunhofer did not understand the significance of these "spectral" lines. However, Gustav Kirchoff (1824-1887) and Robert Bunsen (1811-1899) verified that the lines seen in spectra were characteristic of different elements. These spectral lines can be used as a "fingerprint" to identify an element whenever it is present. The lines in the solar spectrum seen by Fraunhofer indicate the presence of 65 elements. The television program describes these lines that cross the light spectra from the sun and stars. These lines indicate the compositions of stars to the astronomer.

Scientists later discovered that visible light is but a small fraction of the radiant energy in the universe. Known collectively as electromagnetic radiation, the energy ranges from radio and infrared waves, through the visible spectrum of light, to ultraviolet light, and X- and gamma rays.

This lesson describes how the values for the speed of light were calculated and how the wavelengths and frequencies of light and other kinds of electromagnetic radiation are determined and measured. The lesson also explains how light energy is propagated and dissipated according to the inverse-square law.

The properties and behavior of light are included in this lesson as well. You will be introduced to the two prevailing theories explaining the nature of light — the wave theory and the particle, or corpuscular, theory. In actuality, both the wave and particle theories are required to explain the nature and behavior of light. An explanation of this combined, or quantum, theory will add to your understanding of light. A brief review of the structure of atoms, combined with all the other information about light, contributes to this lesson's explanation of how astronomers and other scientists are able to know so much about the stars and about the clouds of gas and dust in between the stars when all we see of them here on Earth are tiny pinpricks of light.

Learning Objectives

When you have completed all assignments in this lesson, you should be able to:

1. Express the interrelationship between the speed of light, wavelength, and frequency.

2. List the major regions of the electromagnetic spectrum in order.

3. Name sources of each kind of electromagnetic radiation.

4. Describe, in general terms, the wave theory, particle theory, and quantum theory of electromagnetic radiation.

5. Describe the inverse-square law of electromagnetic radiation.

6. Describe techniques used by Roemer, Foucault, and Michelson to determine the speed of light.

Assignments for This Lesson

Before Viewing the Program

Read the overview at the beginning of this lesson and look over the learning objectives that precede the assignments.

Read pages 69-79 and 84-85 of Chapter 4 in the text.

View Program 4, "Electromagnetic Radiation"

As you view the program, look for:

- the demonstrations showing both the wave and the particle characteristics of light.

- the diagrams and other visuals that illustrate the electromagnetic spectrum and show visible light as a small portion of that spectrum.

- the explanations of the terms "frequency," "wavelength," "speed," and "energy level" as they relate to various portions of the electromagnetic spectrum and how these properties are interrelated.

- the description of the nature of radio and infrared astronomy as important complements to astronomy using optical light.

- the synthesis of the particle and wave theories of light into a modern quantum theory that recognizes the fundamental duality of all matter.

After Viewing the Program

Review the text assignments.

Review the terms that appear in italics or boldfaced type in the text. In particular, note the key terms and concepts listed in the study activities section.

Complete the study activities and the self-test in this study guide; then check your answers with the answer key at the end of the lesson.

Complete the following exercises: 2, 3, 4, and 5 on pages 85-86 in text Chapter 4.

Study Activities

Key Terms and Concepts

electromagnetic radiation
cosmic ray
solar wind
speed of light
wavelength
radio wave
infrared radiation
(visible) light
ultraviolet radiation
X-rays
gamma rays
microwave radiation
photons
inverse-square law of light
angstrom
electromagnetic spectrum
frequency
Planck's constant
dispersion
spectroscope
spectrograph
Fraunhofer lines
quantum theory (mechanics)

Using What You've Learned

1. Two light sources side by side emit light of the following frequencies:
   a. 6.00×10^{14} cycles/second (Hertz) and
   b. 5.00×10^{14} cycles/second (Hertz)

      1) Which appears more red?
      2) What is the wavelength of light from b?
      3) Which of the two beams will be deviated the least in passing through a prism?
      4) Which light has the most energetic photons?

2. A 100-watt lamp is observed at a distance of 200 meters. The lamp is moved closer so that it is only 50 meters away. It now just matches in apparent brightness a lamp that is known to be 100 meters away. What is the power of the second lamp?

Recommended Readings

For supplementary reading see the bibliography in the back of this study guide.

Self-Test

(Select the one best answer.)

Objective 1: Express the interrelationship between the speed of light, wavelength, and frequency.

1. Use the equation $c = \lambda f$ to determine the wavelength (λ) of light at a frequency (f) of 6.666×10^{14} cycles/second. The speed of light is 3×10^8 meters/second.
   a. 4.5×10^{-7} meters
   b. .0006 meters
   c. 2.5×10^5 meters
   d. 5,000 meters

Objective 2: List the major regions of the electromagnetic spectrum in order.

2. Select the sequence that lists the forms of electromagnetic energy in order of decreasing wavelength and increasing energy.
   a. radio, microwave, infrared, visible, ultraviolet, X-ray, gamma ray
   b. gamma ray, X-ray, ultraviolet, visible, infrared, microwave, radio
   c. visible, infrared, ultraviolet, microwave, X-ray, radio, gamma ray

3. Electromagnetic radiation can be heard by the human ear.
   a. true
   b. false

Objective 3: Name sources of each kind of electromagnetic radiation.

4. The types of radiation emitted by the sun are
   a. infrared.
   b. visible.
   c. X-ray.
   d. ultraviolet.
   e. radio.
   f. all of the above.

5. A very cool cloud of hydrogen gas would be observed to radiate most of its energy as
   a. X-rays.
   b. ultraviolet radiation.
   c. infrared radiation.
   d. visible light.

Objective 4: Describe, in general terms, the wave theory, particle theory, and quantum theory of electromagnetic radiation.

6. Electromagnetic radiation assumed to travel in tiny quanta of energy called photons describes the
   a. particle theory of light.
   b. quantum theory of electromagnetic radiation.
   c. wave theory of light.
   d. wavicle theory of light and energy.

7. Refraction describes light as it
   a. bends when passing from one substance and into another.
   b. breaks up into the colors of the rainbow.
   c. bounces off a smooth, shiny surface.
   d. produces a spectrum of colors.

8. Electrons absorb incoming photons, thus raising the electron to higher energy levels. A photon is emitted when the electrons lose energy. This process describes how the atom
   a. destroys light.
   b. absorbs and emits light.
   c. changes its characteristics.
   d. creates light.

Objective 5: Describe the inverse-square law of electromagnetic radiation.

9. If an object is brought three times closer, it would appear ______________ times ______________.
   a. three, fainter
   b. six, fainter
   c. nine, fainter
   d. three, brighter
   e. nine, brighter

Objective 6: Describe techniques used by Roemer, Foucault, and Michelson to determine the speed of light.

10. Olaus Roemer determined the speed of light using
    a. mirrors and an electric motor.
    b. a rotating disk with many notches in it.
    c. the satellites in orbit around Jupiter.
    d. a rotating mirror located at the Mt. Wilson Observatory.

Answer Key

1. a	6. b
2. a	7. a
3. b	8. b
4. f	9. e
5. c	10. c

Astronomical Observations 5

Overview

When Galileo first viewed the planet Saturn through his telescope, he was startled to see protuberances extending from the planet. In his sketches, it appeared as if Saturn had "ears" or two large satellites, one on either side. Because of the greater magnifying power and better resolution of image, later telescopes showed these bulges were actually a large ring system around the planet. Study of the planets and stars has advanced through technology, as scientists have devised new instruments to gather information from distant objects.

However, you learned in Lesson Four that advances in astronomy have also depended upon better understanding of natural laws. The riddle of spectrum lines caused by gases was solved only when Niels Bohr developed a new model of the atom — that smallest unit of matter — which predicted the release of energy at certain specific levels, manifesting itself as lines on specific portions of the spectrum. Scientists now use these lines in the spectra of stars not only to analyze the composition of a star and the gases surrounding it, but to identify the states of the atoms making up that composition. The Doppler effect, that shift in the lines of a spectrum that indicates movement of a light source toward or away from Earth, has been employed to identify double stars and describe the motions within Saturn's rings.

In this lesson you'll be learning about some of the techniques and equipment that astronomers use to make their discoveries. It's pretty evident that the secrets of outer space don't reveal themselves to the casual viewer, even if he or she is looking through a telescope. You'll be reminded of the universality of nature's laws as the meaning of some of the information gained from spectroscopy is explained to you. You'll also be studying some of the technical features of telescopes and be learning about techniques that are used to overcome problems of reflected and refracted light. Your text will describe several instruments, like the radio telescope, that are used to explore portions of the electromagnetic spectrum that aren't visible to us.

In the television program, you'll learn about the Infra-Red Astronomical Satellite (IRAS), which studied deep infrared radiation from a point high above Earth. (Earth's atmosphere blocks most of the radiation at these energy levels from reaching the ground.) On August 9, 1983, the Jet Propulsion Laboratory announced the discovery by the IRAS of a disk of large-sized particles around the star Vega (in the constellation of Lyra, the harp). These particles may be the size of planets. If this is indeed the case, then humankind has seen the first solar system (possibly in the process of forming) around another star, or merely a large belt of asteroid-sized objects. IRAS, incidentally, was partially credited with discovering a hitherto-unknown comet (IRAS-Araki-Alcock) in the spring of 1983.

The television program concludes with a brief tour of the planets of the solar system. You'll find this a good prologue to the next seven lessons, in which you'll explore each of the planets in much more detail.

Before you complete the assignments for this lesson, you will have explored some aspects of several branches of science, including optics, atomic physics, and electronics. And you'll probably agree that one has to understand how astronomers study the universe in order to adequately appreciate the meaning of their discoveries.

Learning Objectives

When you have completed all assignments in this lesson, you should be able to:

1. Describe the fundamental principles involved in astronomical spectroscopy including concepts of energy and atomic structure.

2. Name and describe the three basic types of spectra and explain how each is formed.

3. Understand the Doppler effect.

4. Define and explain the use of Planck's law, Wien's law, and the Stefan-Boltzmann law.

5. Identify the various telescopes used by astronomers and list the advantages and disadvantages of each.

6. Identify and describe the devices used by astronomers to study nonvisible electromagnetic radiation.

7. Describe how the atmosphere influences astronomical observations.

Assignments for This Lesson

Before Viewing the Program

Read the overview at the beginning of this lesson and look over the learning objectives that precede the assignments.

Read pages 75-85 of Chapter 4 and all of Chapters 5 and 6 in the text.

View Program 5, "Astronomical Observations"

As you view the program, look for:

- the explanation of spectroscopy.
- the illustration of how elements are identified.
- the explanation of the formation of continuous, bright line, and dark line spectra (Kirchoff's laws).
- the explanation of the Doppler effect.
- a description of the atomic structure and how it relates to the production of spectra.
- some of the instruments used to study the nonvisible portions of the electromagnetic energy spectrum.
- a description of the development and expectation of the Space Telescope.

After Viewing the Program

Review the text assignments.

Review the terms that appear in italics or boldfaced type in the text. In particular, note the key terms and concepts listed in the study activities section.

Complete the study activities and the self-test in this study guide; then check your answers with the answer key at the end of the lesson.

After much experimentation in his laboratory, Gustav Kirchoff (1824-1887) was able to formulate three laws that described the production of spectra. These laws are empirical; in other words, they are based completely on observation and do not explain why or how spectra are produced. They only describe the conditions under which each of the spectra appears. Kirchoff's three laws are:

1. Continuous spectrum — produced by a glowing solid, liquid, or a gas under very high pressure (uninterrupted sequence of colors of the rainbow).

2. Bright line spectrum — produced by a rarefied (thin), glowing gas (a series of discrete, brightly colored lines).

3. Dark line spectrum — produced when the light from a source producing a continuous spectrum passes through a cooler, rarefied gas (a series of discrete dark lines superimposed on a bright continuous background. These dark lines are wavelengths that have been removed from the light of the continuous source and are characteristic of the composition of the cooler gas).

Complete the following exercises: 6, 8, and 9 on page 86 in text Chapter 4, exercises 1 and 4 on page 104 in text Chapter 5, and exercises 1 and 6 on page 119 in text Chapter 6.

Study Activities

Key Terms and Concepts

Planck's law
Wien's law
Stefan-Boltzmann law
black body radiation
Doppler effect (shift)
continuous spectrum
emission spectrum (bright line)
absorption spectrum (dark line)
spectroscopy
energy levels (orbits)
Balmer lines
atomic transitions
ionization
atmospheric windows
refracting telescope
reflecting telescope
photometry
interferometry
cosmic rays
radio telescope
Space Telescope

Using What You've Learned

1. Describe the three types of spectra produced and explain how each is obtained. For additional information, find another astronomy text and read about Kirchoff's laws.

2. The drawing here shows a hydrogen atom with several energy levels; the atom has one proton and one electron.

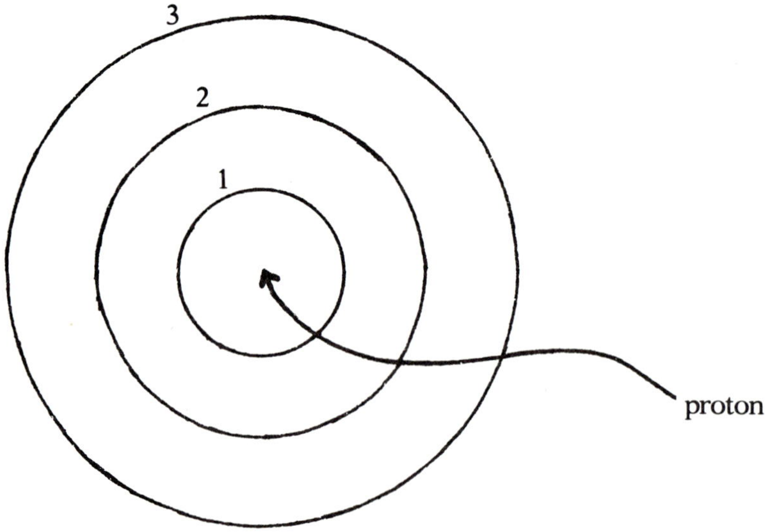

Figure 5.1. Hydrogen atom.

a. In which orbit or level is the electron most likely to be found?
b. If the atom were to absorb a photon to raise the electron to a higher energy level, would it require more energy to raise the electron from (1) to (2) or from (1) to (3)?
c. Which of the two changes in (b) requires the photon of longer wavelength?
d. Which change requires the photon of higher frequency?

3. Listed are data for two telescopes. Using the data given, try to answer the questions that follow.

Type of Telescope	Reflector	Refractor
Diameter of objective	240 cm	75 cm
Focal length of objective	900 cm	1,200 cm
Focal length of eyepiece	5 cm	1.5 cm

a. Which has the greatest magnification?
b. Which has the greatest resolving power?
c. Which has the greatest light-gathering power?
d. Which is best for photographing a faint cluster of galaxies?
e. Which uses a lens as the objective rather than a mirror?

Recommended Readings

For supplementary reading see the bibliography in the back of this study guide.

Self-Test

(Select the one best answer.)

Objective 1: Describe the fundamental principles involved in astronomical spectroscopy including concepts of energy and atomic structure.

1. A bright line or emission spectrum is produced when electrons move from orbits of
   a. higher energy to orbits of lower energy.
   b. lower energy to orbits of higher energy.
   c. equal energy.

Objective 2: Name and describe the three basic types of spectra and explain how each is formed.

2. A hot, rarefied gas produces a(n)
   a. dark line (absorption) spectrum.
   b. continuous spectrum.
   c. ultraviolet-dark line spectrum.
   d. bright line (emission) spectrum.

3. A dark line (absorption) spectrum would be produced by a
   a. glowing cloud of hot hydrogen.
   b. cool gas cloud of nitrogen.
   c. star.
   d. glowing cloud of hot hydrgen behind a cloud of cool nitrogen gas.

Objective 3: Understand the Doppler effect.

4. If a star is receding from an observer, its spectral lines would appear to be
   a. shifted toward the blue end of the spectrum.
   b. shifted toward the red end of the spectrum.
   c. unchanged.
   d. changed only in the red wavelength.

Objective 4: Define and explain the use of Planck's law, Wien's law, and the Stefan-Boltzmann law.

5. According to the Stefan-Boltzmann law, two stars exactly the same size but one twice as hot as the other would emit their total energy in a ratio of
   a. 2 to 1.
   b. 4 to 1.
   c. 8 to 1.
   d. 16 to 1.

Objective 5: Identify the various telescopes used by astronomers and list the advantages and disadvantages of each.

6. The 40-inch telescope located at the Yerkes Observatory is a
   a. refracting telescope.
   b. reflecting telescope.
   c. Cassegrain.
   d. Schmidt camera.

7. One of the advantages of the Space Telescope is that it is
   a. the largest telescope ever built.
   b. unaffected by the turbulence in Earth's atmosphere.
   c. unaffected by temperature.
   d. the first telescope to be controlled by computers.

8. The fact that shorter (blue) wavelengths of light are refracted more than longer (red) wavelengths explains why light is
   a. reflected off parabolic mirrors.
   b. refracted off spherical mirrors.
   c. separated or dispersed by a prism.
   d. absorbed by glass or opaque materials.

Objective 6: Identify and describe the devices used by astronomers to study nonvisible electromagnetic radiation.

9. Several radio telescopes placed far apart but linked electronically and by computer enable the astronomer to observe objects with greater resolving power. This technique is called
   a. spectroscopy.
   b. optical astronomy.
   c. interferometry.
   d. radiation analysis.

10. The Einstein satellite was designed to detect
    a. gamma rays.
    b. ultraviolet radiation.
    c. infrared radiation.
    d. X-rays.

Objective 7: Describe how the atmosphere influences astronomical observations.

11. Visible and radio wavelengths reach the surface of Earth. It is said that these radiations pass through
    a. radiation belts.
    b. atomic nuclei.
    c. atmospheric windows.
    d. thin clouds or haze.

Answer Key

1. a
2. d
3. c
4. b
5. d
6. a
7. b
8. c
9. c
10. d
11. c

Earth — The Water Planet

6

Overview

You may feel it's a bit odd to begin a detailed study of the solar system with Earth. Those great minds of astronomy — Copernicus, Galileo, Kepler, and Newton — long ago demonstrated that Earth is only one part of a huge family (if you include small bodies like the asteroids and even meteor-sized particles) circling the sun. However, there are several important reasons for looking closely at Earth early in the course. For one thing, Earth is the astronomical structure that science knows most about through firsthand evidence. It's a product of the same natural forces responsible for all the structures in the universe. Dr. Krupp will point out to you in the television program that astronomers need to know about the structure and processes of Earth as a basis for comparison with other similar bodies. Moreover, Earth is the "observation platform" most often used for gathering information about the rest of the universe. It's true that quite recently our scientists have begun to gather a little firsthand data from other planets: humans have walked on the moon and brought mineral samples home for study, and instrument packages have landed on Mars and Venus. Space telescopes and space probes also have joined the arsenal of astronomical devices. Even with these fantastic advances, most of the knowledge humankind has gained about the universe has come from observations "at home," here on Earth.

In this lesson, you'll learn some important physical characteristics of Earth, such as its size, shape, and mass. You'll also learn about the science that has revealed many of the secrets of Earth's interior, and you'll "view" a cross-section of Earth as science understands it today. You'll take a look at Earth's surface and atmosphere as well. If you've seen pictures of Earth taken from space, it won't surprise you to learn that three-fourths of the planet is covered with water. The television program and the text describe the "hydrosphere" and the atmosphere and, in the television program, you'll see some beautiful films of the aurora, the "Northern lights." These strange phenomena are, in part, an effect of Earth's magnetic field, or "magnetosphere," which attracts and directs particles from the sun toward Earth's magnetic poles.

One of the topics in this lesson might leave you with an insecure feeling that you're taking a journey on a raft, rather than living on a firmly secure continent. The continents, according to the current widely accepted model of Earth, ride atop huge crustal plates which in turn "float" on a semifluid interior. Of course, the actual movement of the plates and their continents is slow, measurable only over geologic time. Earthquakes and volcanic activity, however, indicate areas where these plates scrape against or collide with each other. One bit of suggestive evidence you can easily examine on a world map or globe is the shape of South America's eastern coast compared with Africa's western one: the two almost fit together like pieces in a very simple jigsaw puzzle. In fact, evidence from natural records such as fossils and magnetic orientation of rocks indicates that all of Earth's land masses once formed a single huge "supercontinent" eons ago. Further information about the crustal plates and their movement has been gathered by photographing Earth from space, as you'll see in the television program.

This lesson also includes a review of Earth's movements through space. This Earth, which seems so firm and solid for us, is actually moving in at least four different directions at the moment you read this. Can you name or describe all four?

Learning Objectives

When you have completed all assignments in this lesson, you should be able to:

1. State the size, shape, and mass of Earth and describe its internal structure.

2. Describe the Foucault experiment proving that Earth rotates on its axis.

3. Describe how the tilt of Earth's axis gives rise to the seasons.
4. Describe the properties of Earth's atmosphere.
5. Describe the nature of Earth's magnetic field.
6. Explain the theory of plate tectonics and continental drift.

Assignments for This Lesson

Before Viewing the Program

Read the overview at the beginning of this lesson and look over the learning objectives that precede the assignments.

Read the introduction to Part II in the text, pages 120-125. Read Chapter 7 in the text.

Reread carefully the text section "Tides" in Chapter 2, pages 44-48.

Reread carefully the text sections "The Foucault Pendulum," pages 53-54, and "The Seasons," pages 54-56, in Chapter 3.

View Program 6, "Earth: The Water Planet"

As you view the program, look for:

- a discussion of the reasons why Earth is called the water planet and an explanation of the hydrological cycle.
- a description and an explanation of the levels of Earth's atmosphere.
- the Foucault pendulum at Griffith Observatory demonstrating that Earth rotates on its axis.
- an explanation of precession of Earth.

After Viewing the Program

Review the text assignments.

Review the terms that appear in italics or boldfaced type in the text. In particular, note the key terms and concepts listed in the study activities section.

Complete the study activities and the self-test in this study guide; then check your answers with the answer key at the end of the lesson.

Complete the following exercises: 3, 4, 8, and 10 on page 144 in text Chapter 7.

Study Activities

Key Terms and Concepts

troposphere
stratosphere
mesosphere
thermosphere
exosphere
ionosphere
magnetosphere
Van Allen layer
core
mantle
crust
plate tectonics
continental drift
lithosphere
asthenosphere
Pangaea

Using What You've Learned

1. Diagram the internal structure of Earth, identifying the four major layers. Continue the diagram to include the five major regions of Earth's atmosphere. List the altitude of each region and determine where interplanetary space begins.

2. Answer the following questions:
   a. Why is the sky blue?
   b. Why does the sun sometimes appear larger at the horizon, just before sundown, than when it is overhead?
   c. How do halos form around the sun and the moon?

3. Visit the geology and earth science exhibits at your local museum to learn more about Earth.

4. You discover a new satellite of Earth. You send radar waves to it and receive the radar echo exactly two seconds after the signal is transmitted. What is the distance to the satellite in kilometers?

Recommended Readings

For supplementary reading see the bibliography in the back of this study guide.

Self-Test

(Select the one best answer.)

Objective 1: State the size, shape, and mass of Earth and describe its internal structure.

1. The rapid rotation of a non-rigid planet causes the
   a. oblateness of Earth.
   b. auroras (Northern lights).
   c. continents of Earth.
   d. orbital lock of the moon's rotation.

2. The layers of Earth, in order from the surface to the center of Earth, are
   a. crust, liquid inner core, mantle, core.
   b. crust, mantle, liquid outer core, solid inner core.
   c. mantle, crust, solid core, liquid inner core.
   d. solid core, liquid outer core, mantle, crust.

Objective 2: Describe the Foucault experiment proving that Earth rotates on its axis.

3. The daily rising and setting of the sun and stars, the behavior of the Foucault pendulum, and the ocean and wind currents are a direct result of Earth's
   a. magnetic field.
   b. rotation.
   c. gravity.
   d. atmospheric composition.

Objective 3: Describe how the tilt of Earth's axis gives rise to the seasons.

4. The 23½° inclination (obliquity between the ecliptic plane and Earth's equator) results in Earth's
   a. magnetic field.
   b. gravity.
   c. rotation.
   d. atmospheric conditions.
   e. seasons.

Objective 4: Describe the properties of Earth's atmosphere.

5. The most abundant (78%) element in Earth's atmosphere is
   a. nitrogen.
   b. oxygen.
   c. carbon dioxide.
   d. water.

6. The oxygen in Earth's atmosphere is believed to have resulted from
   a. volcanism.
   b. dissociation of water by ultraviolet radiation from the sun.
   c. photosynthesis by plant life.
   d. none of the above.

Objective 5: Describe the nature of Earth's magnetic field.

7. A doughnut-shaped region surrounding Earth, where many charged particles are trapped, is called the
   a. magnetosphere.
   b. Van Allen layer.
   c. aurora borealis.
   d. magnetic field.

8. The region around Earth, or around another planet, occupied by its magnetic field is called the
   a. magnetometer.
   b. Van Allen layer.
   c. Ionization Zone.
   d. magnetosphere.

Objective 6: Explain the theory of plate tectonics and continental drift.

9. Earth's crust is made up of giant sections called plates. What is the motion of these sections over the underlying mantle of Earth called?
   a. plate movement
   b. plate tectonics
   c. crustal block movement
   d. continental block movement

10. It is believed that all of the present-day continents were at one time connected as a single giant continent. Scientists have given this continent what name?
    a. Atlantis
    b. Ionia
    c. Pangaea
    d. Argonia

Answer Key

1. a	6. c
2. b	7. b
3. b	8. d
4. e	9. b
5. a	10. c

Lunar Geology 7

Overview

Human beings have been to the moon. They have even brought pieces of it back. Since July 1969, when two Apollo astronauts first walked on the surface of the moon and collected soil samples, lunar geology has been rapidly moving from the realm of conjecture to the realm of fact. Since then, six manned Apollo and at least four unmanned Soviet Luna missions have brought over 1,000 pounds of "moon rocks" to Earth. These lunar samples are being studied and restudied, as are thousands of photographs of the moon's surface. Many of the instruments left on the surface or placed in circumlunar orbit by these and other Soviet and American programs to detect and record seismic, magnetic, and other phenomena are still operating and providing earthbound scientists with new data. Oddly enough, the results of these lunar experiments sometimes have raised more questions about the makeup of the moon and the origin of the Earth-moon system than they have answered. This is what continues to make the study of lunar geology interesting and, perhaps, even more intriguing now than before Earth's nearest neighbor was reached by humans.

This lesson in lunar geology begins with a description of the Earth-moon system and reviews the gross dimensions of the moon: its size, mass, and distance from Earth. So large is this satellite in relation to its primary that Earth and the moon are sometimes referred to as a "double planet." In this lesson, various motions of Earth and moon, particularly their motion around their center of mass (barycenter), are described, as well as the observation that the moon is pulling away from Earth while the rotation of Earth itself slows down.

This lesson considers the great advancements in knowledge about the moon gained since the advent of the lunar exploration programs, beginning with the Pioneer, Ranger, and Lunik probes, followed by Surveyor and Luna and, in particular, the manned Apollo missions. The general appearance of the moon's surface and conditions that exist on that surface are described and examined. You'll also learn how previous conjectures about the age of the moon were generally confirmed by various lunar missions. However, suppositions about the structure of the moon's interior — once believed to be homogenous — were drastically altered by these same missions.

We now know that the moon has a "differentiated" structure, with a crust, a mantle, and a core. Without water or atmosphere — and consequently, without the erosion of wind or weather — the moon's surface presents a far more accurate story of that body's origin and development than Earth reveals about its own origin. The three or four basic types of lunar material are described and their origins pondered. One type of soil is ancient, crustal, and predominates in the light-colored "highlands" region of the moon. This area is fractured and full of craters, testifying to the impact of long-ago meteors. Another soil type, darker and lava-like, covers the plains of "maria" of the moon. Its presence suggests that the moon once had a heated core and experienced volcanism. Still another soil type is material that was probably excavated from the moon's old crust by enormous meteor impacts and pulverized into a grainy and powdery soil.

This lesson describes three theories about the origin of the moon — the fission, the capture, and the binary accretion theories — and details the evidence currently favoring the accretion theory. The lesson also outlines a scenario by which the Earth-moon system may have been formed and the moon may have acquired its present appearance, so very different from its partner planet. You will discover that while much has been learned, the moon still retains many more mysteries for later generations of scientists to unravel.

Learning Objectives

When you have completed all assignments in this lesson, you should be able to:

1. Describe the various types of lunar land forms.

2. Describe the lunar environment, including temperature variations, gravity, and atmosphere.

3. Name and describe important NASA and Soviet lunar missions, their purposes, and their accomplishments.

4. Describe the major geological findings concerning the lunar surface and its interior.

5. Describe the modern theories of the formation of the moon.

6. Discuss the processes involved in crater formation.

Assignments for This Lesson

Before Viewing the Program

Read the overview at the beginning of this lesson and look over the learning objectives that precede the assignments.

Read pages 145-158 of Chapter 8 in the text and pages 250-251 of text Chapter 13.

View Program 7, "Lunar Geology"

As you view the program, look for:

- the videotaped highlights of the first Apollo moon landing and walk by astronauts Armstrong and Aldrin.

- the informative and detailed visuals illustrating how the moon's surface features (highlands, maria, and craters) may have been formed.

- the description and discussion of three theories of the origin of the moon.

- animation illustrating how the moon might have been formed by the process of binary accretion.

After Viewing the Program

Review the text assignment.

Review the terms that appear in italics and boldfaced type in the text. In particular, note the key terms and concepts listed in the study activities section.

Complete the study activities and the self-test in this study guide; then check your answers with the answer key at the end of the lesson.

Complete the following exercises: 1, 2, 3, 5, and 6 on page 162 in text Chapter 8.

Study Activities

Key Terms and Concepts

Ranger
Surveyor
orbiter
Lunokhod
Apollo
mare (maria)
rilles
craters
highlands (lunar)

Using What You've Learned

1. List the similarities and differences between lunar and Earth crustal rocks. Use sources other than your text for additional information.

2. Describe the differences between the near and far sides of the moon. Compare the number of maria, highlands, craters, mountain ranges, lava flooding, and thickness of the lunar crust.

3. Draw a diagram illustrating how you might measure the height of a lunar mountain from the length of its shadow. Describe your procedure and, in each case, explain what must be known or assumed.

Recommended Readings

For supplementary reading see the bibliography in the back of this study guide.

Self-Test

(Select the one best answer.)

Objective 1: Describe the various types of lunar land forms.

1. Most lunar craters were probably formed by
   a. impacts.
   b. volcanism.
   c. collapse of old material.
   d. unknown processes.

2. Low regions and ancient basins filled by lava flows describe what lunar feature?
   a. craters
   b. lunar maria
   c. rilles
   d. highlands

Objective 2: Describe the lunar environment, including temperature variations, gravity, and atmosphere.

3. If you weighed 150 pounds on Earth you would weigh __________ pounds on the moon.
   a. 25
   b. 50
   c. 150
   d. 300

4. Stars would be visible during the day on the moon because
   a. there is no atmosphere to scatter sunlight.
   b. the moon is always in darkness
   c. the moon always points away from the sun.
   d. This statement is false. Stars are never seen during the daytime.

5. One of the reasons for the extreme temperatures on the lunar surface ($^{-}170°$ C to $^{+}130°$ C) is that the moon
   a. is closer to the sun than Earth is.
   b. is smaller than Earth.
   c. has no atmosphere.
   d. has no magnetic field.

Objective 3: Name and describe important NASA and Soviet lunar missions, their purposes, and their accomplishments.

6. Which of the following lunar exploration missions had the capability of roaming over the lunar surface?
   a. Surveyor
   b. Lunokhod
   c. Ranger
   d. lunar orbiter

7. The most striking difference between the near and far sides of the moon is that
   a. the far side has fewer maria than the near side.
   b. there are no craters on the far side.
   c. the far side is in constant darkness.
   d. All of the above are true.

Objective 4: Describe the major geological findings concerning the lunar surface and its interior.

8. A chemical analysis of the lunar maria rocks indicates that they are
   a. identical to Earth rocks in every way.
   b. similar to volcanic basalts of Earth.
   c. completely different from Earth rocks in every way.

9. Evidence for a differentiated moon has been acquired from
   a. the lunar samples returned to Earth.
   b. recordings made by seismographs left on the lunar surface by Apollo astronauts.
   c. samples of lunar material taken from deep beneath the surface.
   d. none of the above.

Objective 5: Describe the modern theories of the formation of the moon.

10. The most widely accepted theory to explain the formation of the moon states that the moon was formed
    a. elsewhere and was captured by Earth.
    b. as part of Earth and broke off the newly formed planet.
    c. simultaneously with Earth out of the same nebula — a twin planet.

Objective 6: Describe the processes involved in crater formation.

11. The bright rays seen around some craters were caused by
    a. lava flows from nearby volcanos.
    b. asteroid material showering down from space before the actual impact of the object.
    c. lava flows from cracks in the lunar surface produced by meteor impacts.
    d. material ejected out of the crater as the result of a high-speed asteroid impact.

Answer Key

1. a
2. b
3. a
4. a
5. c
6. b
7. a
8. b
9. b
10. c
11. d

7/11

Mercury and Venus

8

Overview

In the evening or morning sky during much of the year, one object, Venus, far outshines all others except our moon. At times, Venus is visible for about three hours after sunset or before sunrise. In fact, Venus at its brightest may be seen in full daylight if the air is especially clear. During World War II, Venus was occasionally spotted during daylight and mistaken for an "enemy balloon"; even today it is sometimes reported as an "unidentified flying object." Tiny Mercury is visible, like Venus, but only after sunset or before sunrise and for even briefer periods of time. Although it can be one of the brighter lights in the evening or predawn sky, Mercury is partly obscured by twilight during most of the hours that it's visible. A light haze or dust can make Mercury invisible to the unaided eye.

Through even a small telescope, Venus reveals the first of its surprises. At its brightest, it will appear like a smaller version of the moon, without visible markings, between new and first quarter phase. Mercury, too, will reveal phases like the moon, although a much stronger telescope would be needed to see its partly lighted disk.

Why do these two planets remain so close to the sun and display partly lighted disks? Because they both go around the sun inside Earth's orbit and therefore are always closer to the sun than Earth. The phases are caused by the relative positions of Earth, Venus or Mercury, and the sun — similar to the phases of the moon as observed from Earth during the month. An animated diagram in the television program will help you relate the positions of these bodies to their apparent movements seen from Earth and may help you visualize various phases in relation to specific positions.

Some amateur astronomers enjoy seeing how early or how late they can spot Venus and Mercury with the naked eye in the morning or evening sky. A current almanac, amateur astronomy magazine, or sometimes even a local newspaper column will help you determine when Mercury or Venus should first be visible. If you'd like to try observing the planets, you might find that a pair of ordinary sports binoculars would help you to see Mercury. If you aren't certain you've seen this elusive planet, two or three nights viewing will confirm your sighting, because it will move much more rapidly across its portion of the sky than the stars and it will be in a noticeably different position at the same time each night.

This lesson also acquaints you with details about Venus and Mercury that were unknown through telescopic astronomy. You'll learn that both are classed as "terrestrial" planets. Terrestrial planets are Earth-like planets with similar densities (hard and rocky) and surface features (mountains, craters, and canyons). This does not mean that these planets have Earth-like temperatures or atmospheres. Mercury surprised astronomers with its appearance, which is amazingly moon-like, and a weak magnetic field similar to that of Earth. Flybys and landings on Venus revealed the extremely high temperature and pressure at the planet's surface and the acid content of its cloud layers. Radar mapping has provided some data on the overall surface of this relatively "smooth" planet. Before the flybys, scientists used radar to discover that the rotation of Venus is retrograde (east to west). But the real surprises came when the spacecraft landings discovered that the amount of light that reaches the surface through the thick clouds is more than expected and that there are more and sharper rocks littering its surface.

If you consider only size and distance from the sun, Venus comes the closest to being Earth's "twin" in the planet family. As you study the details about Venus in this lesson, you'll be asked to compare and contrast many characteristics of this planet with those of Earth. Through many comparisons, scientists hope to learn more about the nature of each body, and, indirectly, about the natural laws governing planet formation and evolution.

Learning Objectives

When you have completed all assignments in this lesson, you should be able to:

1. Explain the differences between terrestrial and Jovian planets.
2. Describe the orbital and rotational characteristics of Mercury.
3. Explain how the Doppler effect is used to determine the rotation of Mercury.
4. Describe the surface and atmospheric characteristics of Mercury.
5. Describe a model for Mercury's interior.
6. List the major findings of the Mariner 10 mission to Mercury.
7. Recognize and describe the naked-eye appearance of Venus in the sky.
8. Describe the orbital and rotational characteristics of Venus.
9. Compare the general similarities in size, mass, and location of the planets Earth and Venus by citing corresponding statistics for the two planets.
10. Describe the characteristics of Venus's atmosphere.
11. Explain the "greenhouse effect" of Venus's atmosphere and how it influences the surface temperature on Venus.
12. Enumerate what is known about the physical surface of Venus and explain how that information is obtained.

Assignments for This Lesson

Before Viewing the Program

Read the overview at the beginning of this lesson and look over the learning objectives that precede the assignments.

Read the sections on Mercury and Venus, pages 158-162, 163-167, and 179-186 in Chapters 8 and 9 of the text.

View Program 8, "Mercury and Venus"

As you view the program, look for:

- an explanation of the differences between terrestrial and Jovian planets.
- the Mariner 10 photographs of Mercury and Venus.
- an explanation of the motion of Mercury around the sun and description of Mercury's surface and interior.
- the description of the characteristics of the atmosphere on Venus and the "greenhouse effect."
- the Venera photographs of the surface of Venus.

After Viewing the Program

Review the text assignments.

Review the terms that appear in italics and boldfaced type in the text. In particular, note the key terms and concepts listed in the study activities section.

Complete the study activities and the self-test in this study guide; then check your answers with the answer key at the end of the lesson.

Complete the following exercises:

1. Compare the similarities and differences between Mercury, the moon, and Venus. Include not only the surface features of the three, but also their internal structures.
2. What problems would humans face if they tried to colonize the planet Mercury?
3. Compare the characteristics of the atmosphere and surface features of Venus with those of Earth. Do you feel that, in some cases, Earth is becoming more like Venus? Give examples.
4. Develop a diagram that shows the occurrence of the phases of Venus. If Venus is currently visible, use a pair of binoculars to observe the planet and determine where on your diagram Venus must be in order for you to observe its current phase.

Study Activities

Key Terms and Concepts

differentiation
greenhouse effect
terrestrial planet
Jovian planet
perihelion
synchronous
aphelion
transit
gravity assist
retrograde motion
radar observations

Using What You've Learned

For the next several lessons you will be asked to locate the five naked-eye planets on the SC1 Equatorial Star Chart (provided for you by your instructor or available at your local bookstore), determine if they are visible at night, and then, using the constellation charts at the back of your text, go outside and try to find the planets.

The SC1 Equatorial Star Chart

The Equatorial Star Chart shows the position of the stars and constellations that are situated within 60° north and south of the celestial equator. The celestial equator, if you recall, simply divides the sky into the northern and southern hemispheres, just as the equator on Earth divides our planet into two hemispheres. In other words, if you stood on Earth's equator and looked up, the celestial equator (represented by the straight horizontal line through the center of the chart) would be directly overhead.

You will also notice the curved line running through the chart. If you look closely, the line runs through the twelve constellations of the zodiac. This line is the ecliptic, the path the sun appears to follow through the sky over the course of a year. The small numbers along the ecliptic are celestial longitudes and will be used to locate the planets. Remember that the orbital planes of the planets lie close to the ecliptic plane and therefore the planets themselves will be located on or near the ecliptic.

Locating the Stars

You have probably already identified several familiar constellations and located some of the brightest stars. There is a fast way of locating the stars: use their coordinates.

The coordinates of stars are given in terms of their right ascension and declination (longitude and latitude, respectively). The declination determines the position of the stars north (+) or south (−) of the celestial equator. If you look along the vertical line through the center of the chart, the line is marked off in degrees. It starts at 0° on the celestial equator and goes to 60° north (+) and 60° south (−). The right ascension is measured in hours, minutes, and seconds. It starts at the same vertical line at 0 hours. Running along the bottom of the chart, the right ascension goes from 0 to 12 hours on the left half of the chart, and from 12 to 24 hours on the right half of the chart. It should help to remember that right ascension increases from 0 through 24 hours to the left on the chart, because right ascension is always measured toward the east.

Using text Appendix 15, the Brightest Stars, locate several stars by using their coordinates of right ascension and declination. Draw an imaginary horizontal line through the declination coordinate on the chart and an imaginary vertical line through the right ascension coordinate. The two lines should meet at the position of the star.

If you want to compare the chart with the appearance of stars in the night sky, hold your chart overhead, with the top of the chart toward the north, while you are facing south. Those stars which are near 0 hours right ascension will be nearly overhead in the early evening during November; those at 6 hours in February; those at 12 hours in May; and those at 18 hours in August.

Locating the Planets

Table 8.1 provides the celestial longitudes of the sun and the five naked-eye planets at 10-day intervals. Select a day (near the present date) and locate the position of the sun on the ecliptic. (Remember that the celestial longitudes are the small numbers near the ecliptic on the chart.) Using the same dates locate the positions of Mercury and Venus. Using the star charts in the back of the text, go outside (at sundown or just before sunrise) and try to observe the two planets. If the planet is to the right of the sun's position on the ecliptic, it will be visible in the predawn sky unless it is too close to the sun. When the planet is to the left of the sun, it will follow the sun in the sky and may be visible after sunset if not too close to the sun.

Year	Date		J.D.	Sun	Mer	Ven	Mar	Jup	Sat
1987	DEC	20	7150	268	267	298	227	20	264
1987	DEC	30	7160	278	282	310	234	20	266
1988	JAN	9	7170	288	299	322	241	20	267
1988	JAN	19	7180	299	315	335	247	21	267
1988	JAN	29	7190	309	327	347	254	22	268
1988	FEB	8	7200	319	326	359	261	24	269
1988	FEB	18	7210	329	315	11	267	26	270
1988	FEB	28	7220	339	314	22	274	28	271
1988	MAR	9	7230	349	322	34	281	30	271
1988	MAR	19	7240	359	334	45	287	32	272
1988	MAR	29	7250	9	349	55	294	35	272
1988	APR	8	7260	19	6	65	301	37	273
1988	APR	18	7270	29	26	74	307	39	273
1988	APR	28	7280	38	48	82	314	42	273
1988	MAY	8	7290	48	67	87	321	44	273
1988	MAY	18	7300	58	80	90	327	47	272
1988	MAY	28	7310	67	86	90	334	49	272
1989	JUN	7	7320	77	85	84	340	52	271
1988	JUN	17	7330	86	79	78	347	54	270
1988	JUN	27	7340	96	77	73	353	56	269
1988	JUL	7	7350	105	84	72	358	58	268
1988	JUL	17	7360	115	97	76	3	60	268
1988	JUL	27	7370	124	116	81	7	61	267
1988	AUG	6	7380	134	138	89	10	63	266
1988	AUG	16	7390	144	157	97	12	64	266
1988	AUG	26	7400	153	174	107	13	65	266
1988	SEP	5	7410	163	188	117	13	66	266
1988	SEP	15	7420	173	199	128	11	66	266
1988	SEP	25	7430	183	207	139	7	66	267
1988	OCT	5	7440	192	205	150	2	66	267
1988	OCT	15	7450	202	194	162	0	66	268
1988	OCT	25	7460	212	194	174	359	65	268
1988	NOV	4	7470	222	206	186	359	63	269
1988	NOV	14	7480	232	222	198	0	62	270
1988	NOV	24	7490	242	238	211	2	60	271
1988	DEC	4	7500	253	254	223	6	59	272
1988	DEC	14	7510	263	270	235	11	57	273
1988	DEC	24	7520	273	286	248	16	57	275
1989	JAN	3	7530	283	301	260	21	56	276
1989	JAN	13	7540	293	312	273	27	56	277
1989	JAN	23	7550	303	308	286	32	56	278
1989	FEB	2	7560	314	297	298	38	56	279
1989	FEB	12	7570	324	298	311	44	57	280
1989	FEB	22	7580	334	307	323	50	58	281
1989	MAR	4	7590	344	320	336	56	59	282
1989	MAR	14	7600	354	335	348	62	60	282
1989	MAR	24	7610	4	353	1	68	62	283
1989	APR	3	7620	14	12	13	74	64	283
1989	APR	13	7630	23	33	25	80	66	284
1989	APR	23	7640	33	52	38	86	68	284
1989	MAY	3	7650	43	64	50	92	70	284
1989	MAY	13	7660	52	67	63	99	72	284
1989	MAY	23	7670	62	63	75	105	75	284
1989	JUN	2	7680	72	58	87	111	77	283
1989	JUN	12	7690	81	59	100	118	79	283
1989	JUN	22	7770	91	68	112	124	82	282
1989	JUL	2	7710	100	83	124	130	84	281
1989	JUL	12	7720	110	102	137	136	86	280
1989	JUL	22	7730	119	124	149	142	89	279
1989	AUG	1	7740	129	144	161	149	91	278
1989	AUG	11	7750	138	160	172	155	92	278
1989	AUG	21	7760	148	175	184	161	94	278
1989	AUG	31	7770	158	185	196	167	96	278
1989	SEP	10	7780	168	191	208	174	97	277
1989	SEP	20	7790	177	187	219	180	99	277
1989	SEP	30	7800	187	177	231	187	100	277
1989	OCT	10	7810	197	179	242	194	100	278
1989	OCT	20	7820	207	192	253	200	101	278
1989	OCT	30	7830	217	209	264	207	101	279
1989	NOV	9	7840	227	226	274	213	101	280
1989	NOV	19	7850	237	242	284	220	100	281
1989	NOV	29	7860	247	258	292	227	99	282
1989	DEC	9	7870	257	273	300	234	98	283
1989	DEC	19	7880	268	287	305	241	97	284
1989	DEC	29	7890	278	296	308	248	95	285
1990	JAN	8	7900	288	290	306	255	93	286
1990	JAN	18	7910	298	280	301	262	92	288
1990	JAN	28	7920	308	284	295	269	91	289
1990	FEB	7	7930	318	294	291	276	91	290
1990	FEB	17	7940	328	307	293	284	90	291
1990	FEB	27	7950	339	322	297	291	91	292
1990	MAR	9	7960	349	340	304	298	91	293
1990	MAR	19	7970	359	359	312	306	91	294
1990	MAR	29	7980	8	19	322	313	92	294
1990	APR	8	7990	18	37	332	321	93	294
1990	APR	18	8000	28	47	342	328	95	295
1990	APR	28	8010	38	46	353	335	96	295
1990	MAY	8	8020	47	40	4	343	98	295
1990	MAY	18	8030	57	37	16	351	100	295
1990	MAY	28	8040	67	42	27	358	102	295
1990	JUN	7	8050	76	53	39	5	104	295
1990	JUN	17	8060	86	68	50	12	106	294
1990	JUN	27	8070	95	88	62	19	109	294
1990	JUL	7	8080	105	111	74	26	111	293
1990	JUL	17	8090	114	130	86	33	113	292
1990	JUL	27	8100	124	147	98	40	115	291
1990	AUG	6	8110	133	160	110	46	118	290
1990	AUG	16	8120	143	170	122	52	120	290
1990	AUG	26	8130	153	174	135	57	122	289
1990	SEP	5	8140	162	168	147	63	124	289
1990	SEP	15	8150	172	160	160	67	126	289
1990	SEP	25	8160	182	164	172	71	127	289
1990	OCT	5	8170	192	179	185	74	129	289
1990	OCT	15	8180	202	197	197	75	130	289
1990	OCT	25	8190	212	214	210	74	131	289
1990	NOV	4	8200	222	230	222	72	132	290
1990	NOV	14	8210	232	245	235	70	133	290
1990	NOV	24	8220	242	260	247	66	133	291
1990	DEC	4	8230	252	273	260	62	133	292
1990	DEC	14	8240	262	281	272	59	133	293
1990	DEC	24	8250	272	273	285	57	132	294
1991	JAN	3	8260	283	264	298	57	132	295
1991	JAN	13	8270	293	269	310	57	131	297
1991	JAN	23	8280	303	281	323	60	129	298
1991	FEB	2	8290	313	294	335	63	128	299
1991	FEB	12	8300	323	310	348	67	126	301
1991	FEB	22	8310	333	327	0	71	125	302
1991	MAR	4	8320	343	345	13	75	124	303
1991	MAR	14	8330	353	5	25	80	124	304
1991	MAR	24	8340	3	22	37	85	123	304
1991	APR	3	8350	13	30	49	91	123	305
1991	APR	13	8360	23	26	61	96	124	306
1991	APR	23	8370	33	19	73	101	125	306
1991	MAY	3	8380	42	18	84	107	125	306
1991	MAY	13	8390	52	25	96	112	126	306
1991	MAY	23	8400	62	38	106	118	128	307
1991	JUN	2	8410	71	54	117	124	129	307
1991	JUN	12	8420	81	74	127	130	131	306
1991	JUN	22	8430	90	97	136	136	133	306
1991	JUL	2	8440	100	117	144	142	134	305
1991	JUL	12	8450	109	133	151	148	136	305
1991	JUL	22	8460	119	146	155	154	139	304
1991	AUG	1	8470	128	154	158	160	141	303
1991	AUG	11	8480	138	156	155	166	143	302
1991	AUG	21	8490	148	149	149	173	146	301
1991	AUG	31	8500	157	142	144	179	148	301
1991	SEP	10	8510	167	149	140	186	150	300
1991	SEP	20	8520	177	165	141	192	152	300
1991	SEP	30	8530	187	184	146	199	154	300
1991	OCT	10	8540	197	201	152	205	156	300

Table 8.1. Celestial longitudes of the sun and planets.

Recommended Readings

For supplementary reading see the bibliography in the back of this study guide.

Self-Test

(Select the one best answer.)

Objective 1: Explain the differences between terrestrial and Jovian planets.

1. Of the planets listed, which one(s) is (are) terrestrial?
   a. Jupiter
   b. Uranus
   c. Venus
   d. Earth
   e. Saturn

Objective 2: Describe the orbital and rotational characteristics of Mercury.

2. For every ______________ rotations on its axis, Mercury undergoes ______________ revolutions around the sun.
   a. three, two
   b. two, three
   c. one, three
   d. five, three

3. If you observe the sun to be at the zenith as you view it from Mercury, how many of Earth's days will elapse before you view the sun at the zenith again?
   a. 44
   b. 59
   c. 88
   d. 176

Objective 3: Explain how the Doppler effect is used to determine the rotation of Mercury.

4. What type of telescope was used to make the first determination of Mercury's rotation rate?
   a. optical
   b. infrared
   c. refractor
   d. radio (radar)

Objective 4: Describe the surface and atmospheric characteristics of Mercury.

5. Due to its weak gravitational field, Mercury has
   a. no magnetic field.
   b. no atmosphere.
   c. many craters.
   d. very high daytime temperatures.

6. Why would Mercury be an unpleasant place for earthlings to live?
   a. too cold at night
   b. too hot during the day
   c. no oxygen
   d. no vegetation
   e. no water
   f. all of the above

7. Mercury has a magnetic field.
   a. true
   b. false

8. There is no evidence of extensive lava flows on Mercury similar to what we see on the moon.
   a. true
   b. false

9. When seen with the naked eye, Mercury appears most similar to
   a. the moon.
   b. a star.
   c. a shooting star or meteor.
   d. Mercury cannot be seen without the use of a telescope.

Objective 5: Describe a model for Mercury's interior.

10. Mercury's high density is due to a
    a. large inner iron core.
    b. strong magnetic field.
    c. liquid core.
    d. semifluid mantle and thin crust.

Objective 6: List the major findings of the Mariner 10 mission to Mercury.

11. The Mariner photographs of Mercury showed the planet to appear most nearly similar to
    a. Venus.
    b. Earth.
    c. the moon.
    d. Mars.

Objective 7: Recognize and describe the naked-eye appearance of Venus in the sky.

12. If Venus were to be observed as an evening star, it would be seen
    a. above the western horizon after sunrise.
    b. above the eastern horizon before sunrise.
    c. near the meridian at midnight.
    d. above the western horizon after sunset.
    e. above the eastern horizon before sunset.

13. Venus has a high albedo, lies relatively close to the sun, and is close to Earth when it appears brightest. These are three reasons why Venus
    a. goes through phases.
    b. appears so bright in the sky.
    c. orbits the sun in a nearly circular orbit.
    d. has a dense atmosphere.

Objective 8: Describe the orbital and rotational characteristics of Venus.

14. Radar observations of Venus determined its
    a. rotation rate.
    b. albedo.
    c. atmospheric composition.
    d. atmospheric density.

Objective 9: Compare the general similarities in size, mass, and location of the planets Earth and Venus by citing corresponding statistics for the two planets.

15. The orbit of Venus is ______________ circular than Earth's, about ______________ the size of Earth's, and nearly in the same plane as the ______________.
    a. less, ½, sun
    b. more ½, orbit of Mars
    c. more, ¾, Earth's orbit
    d. less, ¾, Earth's orbit

16. Which properties of Venus are very similar to those of Earth?
    a. magnetic field strength, atmospheric composition, temperature
    b. temperature, location in the solar system, size
    c. chemical composition of the surface, mass, temperature
    d. mass, size, location in the solar system

Objective 10: Describe the characteristics of Venus's atmosphere.

17. In order of abundance (from the most to the least), the composition of the atmosphere of Venus is
    a. nitrogen, oxygen, carbon dioxide, water vapor.
    b. carbon dioxide, nitrogen, water vapor, oxygen.
    c. water vapor, oxygen, nitrogen, carbon dioxide.
    d. oxygen, water vapor, nitrogen, carbon dioxide.

18. The fact that the cloud layer of Venus consists of sulfuric acid droplets that are placed well above the surface of Venus with clear atmosphere beneath was determined by
    a. optic telescopes.
    b. radio telescopes.
    c. the Pioneer-Venus orbiter.
    d. space probes sent into the atmosphere and that reached the surface.

Objective 11: Explain the "greenhouse effect" of Venus's atmosphere and how it influences the surface temperature on Venus.

19. Incident sunlight trapped under the carbon dioxide atmosphere of Venus describes the ____________ and produces ____________.
    a. greenhouse effect, extremely high surface temperatures
    b. jet stream, high winds
    c. greenhouse effect, extremely bright lights on the surface
    d. Doppler effect, change in wavelength of the incident radiation

Objective 12: Enumerate what is known about the physical surface of Venus and explain how that information is obtained.

20. The thick clouds of Venus prevent
    a. meteorite impacts on the surface.
    b. the development of a magnetic field.
    c. Earth-based telescopic study of surface features.
    d. lightning from occurring in the atmosphere.

21. Our knowledge about the surface of Venus comes from
    a. satellites in orbit around Earth.
    b. Russian and American landers on the surface of Venus.
    c. Earth-based radio telescopes.
    d. satellites with optical telescopes in orbit around Venus.
    e. radar studies by the Pioneer-Venus and Venera orbiters.
    f. all of the above.

Answer Key

1. c, d
2. a
3. d
4. d
5. b
6. f
7. a
8. b
9. b
10. a
11. c
12. d
13. b
14. a
15. c
16. d
17. b
18. d
19. a
20. c
21. b, c, e

Mars 9

Overview

Mars orbits the sun at a distance half again as far as Earth, and its dimensions are only slightly more than half those of our planet. Yet Mars, of the three terrestrial neighbors of Earth, seems to offer the greatest advantages for telescopic astronomy. It is the only planet whose surface can be observed with Earth-based telescopes. By the 1960s, in fact, the period of rotation for Mars had been long since accurately established, its mass derived from its interaction with its two satellites, and its atmosphere roughly analyzed.

Moreover, the aura of romance and mystery surrounding Mars surpassed by far that of Venus and Mercury. Just at the edge of resolution, for only moments at a time, astronomers would catch a glimpse of formations or structures appearing as faint lines. The optical interference our own atmosphere causes hid these faint features most of the time, and they were too elusive to be captured photographically. This lesson about Mars includes a tribute to those astronomers who tried to resolve the mystery of the "canals of Mars," and the television program will show you some brief excerpts from fantasies about "Martians."

Mars can be a beautiful object to the unaided eye. At times of favorable observations (a configuration called opposition), when it is closest to Earth, Mars shines with a brightness greater than Jupiter's. Its red-orange tint leaves the viewer with little doubt why the old Greek and Roman cultures associated it with their gods of war.

The Mariner and Viking programs effectively ended the question about canals on Mars. Indistinct, hazy features drawn on pre-1965 Mars maps gave way to mountain ranges, volcanic caldera, a vast canyon system, craters, and other complex features. Even though the canals did not exist, Mars held a new mystery for astronomers, because there are visible channels that apparently could have been carved out only by the eroding action of rushing water, and large amounts of water, at that. Where did the water come from, and where did it go?

In this lesson you'll learn about Percival Lowell, whose fascination with the possibility of canals and even life on Mars led to a lifetime of achievement in astronomy. Like Venus, Mars offers some interesting comparisons with Earth in its physical characteristics and its seasons. You'll have an opportunity to compare some rather fantastic features found on that planet: the volcano Olympus Mons and a huge canyon system which dwarf any similar features found on Earth. In the television program, you'll see the Viking landers performing biological experiments and you'll learn about the results of these experiments in both program and text. There will also be photographs of the Martian surface and the two tiny satellites, Phobos and Deimos, which have greatly extended scientific understanding of these bodies. This lesson offers the best answers science has at this time to questions about life on Mars.

Of the other three planets closest to the sun, Mars seems to offer the most hospitable environment (and perhaps the only one) for manned exploration some time in the future. The possibility has been seriously discussed. Just think how an expedition, similar to the Apollo missions, might resolve some of today's scientific questions, and pose even more exciting ones.

Learning Objectives

When you have completed all assignments in this lesson, you should be able to:

1. Describe the orbital, rotational, and axis tilt characteristics of Mars and compare them with Earth's.

2. Describe the Earth-based observations of the Martian surface features.

3. Explain why Mars was once thought to be crisscrossed by elaborate canal systems.

4. Describe the properties of the Martian atmosphere.

5. Describe the Martian surface features as observed by the Mariner and Viking spacecraft.

6. Describe the Martian polar ice caps.

7. Describe the characteristics of the two Martian satellites.

8. Describe the results of the major Viking lander experiments.

Assignments for This Lesson

Before Viewing the Program

Read the overview at the beginning of this lesson and look over the learning objectives that precede the assignments.

Reread the section on Mars in text Chapter 9, pages 163-167 and read carefully pages 167-179.

View Program 9, "Mars"

As you view the program, look for:

- an explanation of how our knowledge about Mars developed, especially the work of Giovanni Schiaparelli and Percival Lowell.

- an explanation of the changing colors of Mars that are visible from Earth.

- the description of the photographs taken by Mariner and Viking orbiters (volcanos, great canyons, craters, and dried riverbeds).

- a detailed description of the nature and content of the polar ice caps.

- the description of the Viking search-for-life experiments and the explanation of the results of those experiments.

After Viewing the Program

Review the text assignment.

Review the terms that appear in italics and boldfaced type in the text. In particular, note the key terms and concepts listed in the study activities section.

Complete the study activities and the self-test in this study guide; then check your answers with the answer key at the end of the lesson.

Early telescopic observations of the planet Mars showed changing polar ice caps and changing colors of the features on the Martian surface. These changes were referred to as the "seasons" of Mars. Originally these color changes were explained as vegetation growth caused by the melting ice caps. When the canals were discovered, water from the ice caps was thought to flow along these canals and to aid the growth of vegetation. Later, when the "ice" caps were determined to contain mostly carbon dioxide and not water, the explanation changed again: color changes were no longer attributed to vegetation but to chemical reactions in the soil.

The most recent analysis of the Mariner photographs has lead scientists to yet another conclusion. The latest and most widely accepted explanation of the Martian seasons centers on the Martian winds. These winds, reaching speeds of several hundred miles per hour, shift large quantities of dust on the Martian surface and expose darker underlying rocks. It is these changing color patterns caused by shifts in surface dust, along with the changing size of the polar "ice" caps, to which observers attribute the seasons on Mars.

Complete the following exercises:

1. List the observations made of Mars from Earth that led some scientists to believe there is life on Mars.

2. Examine a color photograph of Earth as seen from space. (You may wish to use one that is larger than those usually printed in textbooks.) Develop a list of observations that would indicate life exists on the planet. Can you conclude that life (don't even try for intelligent life) exists on Earth? (Just as a point of interest: most photographs of Earth from space were taken from distances of less than 50,000 km from the planet, with a few taken from the moon, over 350,000 km away. The closest Earth gets to Mars is about 75,000,000 km.)

3. Name three Earth-like features on the surface of Mars. How do these features compare with similar ones found on Earth (larger, deeper, more numerous, etc.)?

Study Activities

Key Terms and Concepts

Mariner 6, 7
Mariner 9
Olympus Mons
polar ice caps (Martian)
canals (Martian)
Valles Marineris
Viking orbiter
Viking lander
Phobos
Deimos
obliquity
seasons (on Mars)

Using What You've Learned

1. Locate the planet Mars on the SC1 Star Chart. Use the longitudes given in the table in Lesson Eight. Choose a date near the current one and locate the sun, then determine if Mars can be seen in the nighttime sky. If so, go out and observe it.

2. Try to observe Mars with a small telescope. Draw the surface features that you can see. Try to identify these features on a detailed map of Mars.

3. Read a science fiction novel about the planet Mars. For example, read *Princess of Mars* by Edgar Rice Burroughs (1912). Compare how Mars is described in this book with what we know about it today.

Recommended Readings

For supplementary reading see the bibliography in the back of this study guide.

Self-Test

(Select the one best answer.)

Objective 1: Describe the orbital, rotational, and axis tilt characteristics of Mars and compare them with Earth's.

1. Earth-based obervations have revealed that the rotation period of Mars is
   a. nearly equal to that of Earth.
   b. one-half that of the moon.
   c. nearly equal to that of Venus.
   d. twice that of Jupiter.

2. If Earth were at aphelion and Mars were at perihelion and opposition, Mars would be
   a. farthest from Earth.
   b. nearest to Earth.
   c. nearer to Jupiter than to Earth.
   d. nearer to Venus than to Earth.

Objective 2: Describe the Earth-based observations of the Martian surface features.

3. Changes in the color of the surface features of Mars have been interpreted to mean
   a. the polar ice caps are melting.
   b. the shadows of the satellites of Mars can be seen on the surface.
   c. that Mars may go through seasons.
   d. that shadows of the taller features on Mars can be seen on the surface.

4. As seen from Earth, the ice cap in the northern hemisphere of Mars ______________ in winter and ______________ during the summer.
   a. shrinks, increases
   b. increases, shrinks
   c. shows no change, increases
   d. shows no change, shows no change

5. Percival Lowell believed that the changes in color seen on the surface of Mars were due to
   a. changes in the growth of vegetation on the surface.
   b. chemical reactions.
   c. shadows.
   d. optical illusions.

Objectives 2 and 3: Describe the Earth-based observations of the Martian surface features. Explain why Mars was once thought to be crisscrossed by elaborate canal systems.

6. The first astronomer to detect linear features, later called canals, on the surface of Mars was
   a. Percival Lowell.
   b. Galileo.
   c. Asaph Hall.
   d. Giovanni Schiaparelli.

Objective 3: Explain why Mars was once thought to be crisscrossed by elaborate canal systems.

7. The canals of Mars were
   a. seen by Galileo.
   b. optical illusions.
   c. actual surface features photographed by the spacecraft sent to Mars.
   d. detected by radar studies of Mars.

8. Who proposed that the Martian canals were built to carrry irrigation water from the polar ice caps to the equatorial regions?
   a. Galileo
   b. Percival Lowell
   c. Asaph Hall
   d. Giovanni Schiaparelli

Obejctive 4: Describe the properties of the Martian atmosphere.

9. In order of abundance (most to least), the Martian atmosphere consists of
   a. argon, water vapor, carbon dioxide.
   b. nitrogen, carbon dioxide, water vapor.
   c. water vapor, carbon dioxide, nitrogen.
   d. carbon dioxide, nitrogen, water vapor.

Objective 5: Describe the Martian surface features as observed by the Mariner and Viking spacecraft.

10. Which features have been found on both Earth and Mars?
    a. dry river beds
    b. rivers with running water
    c. volcanos
    d. craters
    e. large canyons
    f. erupting volcanos

11. Olympus Mons is
    a. smaller than Mount Everest.
    b. smaller than Mauna Loa.
    c. the largest volcano in the solar system.
    d. erupting.

12. Valles Marineris is ______________ than ______________ on Earth.
    a. 10 times larger, the Amazon River
    b. 100 times smaller, the Nile River
    c. somewhat larger, the Grand Canyon
    d. 10 times longer, the Grand Canyon

13. Shifting dust patterns caused by the Martian winds is the explanation given for the
    a. polar ice caps.
    b. color changes in the Martian surface seen from Earth.
    c. river-like channels.
    d. craters.

14. Evidence of large amounts of water in the Martian past come from photographs of
    a. dried-up river beds
    b. a residual polar ice cap of water ice.
    c. volcanos, a water source.
    d. lakes and rivers of liquid water.
    e. Martian deserts.

15. The Viking and Mariner spacecraft found that the canals of Mars
    a. exist as described by Lowell.
    b. were the result of dust storms.
    c. were shadows cast by the moons Phobos and Deimos.
    d. do not exist.

Objective 6: Describe the Martian polar ice caps.

16. The composition of the Martian polar ice caps is
    a. only carbon dioxide.
    b. water ice.
    c. frozen carbon dioxide (dry ice) and water ice.
    d. methane ice.

Objective 7: Describe the characteristics of the two Martian satellites.

17. Phobos and Deimos are
    a. the moons of Mars.
    b. two large round surface features.
    c. desert regions on the equator.
    d. ice deposits near the polar ice caps.

18. Phobos and Deimos have been observed to have
    a. craters on their surface.
    b. a composition similar to that of some asteroids.
    c. irregular shapes.
    d. strange grooves on their surface.

Objective 8: Describe the results of the major Viking lander experiments.

19. The Viking lander found
    a. life on the surface of Mars.
    b. complex organic molecules on Mars.
    c. a complex chemistry on Mars imitating life.
    d. easily explained common chemical reactions on Mars.

20. The most logical next step in the exploration of the planet Mars is to send
    a. a manned expedition.
    b. a rover, a vehicle capable of moving across the surface to distant points.
    c. an unmanned craft to crash land on the surface.
    d. another spacecraft similar to the Viking, to land at another location.

Answer Key

1. a
2. b
3. c
4. b
5. a
6. d
7. b
8. b
9. d
10. a, c, d, e
11. c
12. d
13. b
14. a, b, c
15. d
16. c
17. a
18. a, b, c
19. c
20. b

Jupiter 10

Overview

Were you to be "official astronomer" on an exploratory mission from some distant star, you might not find too much of interest as your well-equipped vehicle approached a very average star, despite its importance to the daily lives of beings on the planet Earth. Your expertise as an astronomer and your advanced instruments would tell you that this sun isn't alone: there is a cold body (Jupiter) circling it at about a distance of 1,000 times the star's diameter; and, while the cold body is much smaller, its size is about ten percent of the diameter of the star. A time-consuming study of the relative gravities of these two bodies would reveal, though, that the small, cold companion has less than one percent of the mass of its star. Although there's one other similar-sized but less-massive body circling the same star, and a number of even smaller orbiting bits of matter (which complicated the relative-mass studies a little), a quick survey and a few calculations would show that the star's largest companion contains about 70 percent of all the mass in this system that isn't in the star itself. It's pretty clear that this one planet is the only significant companion to the star; all the others are bits of leftover debris from the system's formation.

Of course, since we live inside this star system (and on one of the "leftover" pieces at that) our astronomers are interested in all the matter found here. As you'll see in the program for this lesson, the scientific team monitoring the first craft bound for Jupiter were especially interested in the tiny members of the asteroid belt, fearing they might pose an impenetrable barrier to voyages beyond the inner solar system. Gigantic Jupiter does seem to dominate our system in many ways. Usually, it is the fourth brightest light in Earth's night sky, the second strongest source of radio signals received from space, and the greatest gravitational force among the planets.

In this lesson you'll explore Jupiter itself and learn, as you study the probable model of the planet's interior, how the lightest element in our universe may become a liquid, and even a solid, there. Other specific features of Jupiter that require your attention include its weather patterns, cloud bands, and the Great Red Spot. Jupiter's magnetic field and the associated band of radiation are as imposing as the planet itself, and have wide-ranging effects. Jupiter and its satellites have been described as a "solar system in miniature," and you'll discover that its largest moons, the Galilean satellites, are as distinctive and individual as any of the worlds studied so far in this course.

Once again, you are to have an adventurous spacecraft's view of a planet. In fact, your view may be better than it would be if you were actually aboard Voyager 1 or 2, because computer processing of the Pioneer and Voyager images both makes the pictures clearer and enhances color contrasts. Time-lapse photography lets you see Jupiter's clouds and rotation in faster-than-life motion in the television program.

Learning Objectives

When you have completed all assignments in this lesson, you should be able to:

1. Describe the size and shape of Jupiter.

2. Describe the orbital and rotational characteristics of Jupiter.

3. Describe the Red Spot.

4. Describe the banded structure of Jupiter's atmosphere.

5. Describe the magnetic field associated with Jupiter.

6. Describe the radiation associated with Jupiter.

7. Name the Galilean satellites and describe their characteristics.

8. Describe the modern model of Jupiter's interior and atmosphere.

9. Describe the rings of Jupiter.

Assignments for This Lesson

Before Viewing the Program

Read the overview at the beginning of this lesson and look over the learning objectives that precede the assignments.

Read the section on Jupiter, pages 187-197 and 199-202, in Chapter 10 of the text. Read also pages 203-210 and 214-216 in text Chapter 11.

View Program 10, "Jupiter"

As you view the program, look for:

- the cloud belts of Jupiter and their motion around the planet.

- the structure of Jupiter's interior.

- the magnetic field and radiation belts of Jupiter and their interaction with the moon Io.

- a description of the nature of the Great Red Spot and an explanation for its existence.

- the surface characteristics and internal structures of the four Galilean satellites (Io, Europa, Ganymede, and Callisto).

After Viewing the Program

Review the text assignment.

Review the terms that appear in italics and boldfaced type in the text. In particular, note the key terms and concepts listed in the study activities section.

Complete the study activities and the self-test in this study guide; then check your answers with the answer key at the end of the lesson.

Complete the following exercises:

1. What causes the banded appearance of the atmosphere of Jupiter?

2. How do the magnetosphere and radiation belts of Jupiter compare with Earth's?

3. Briefly describe the features of and explanation for the Great Red Spot.

4. Describe the surface features of the four Galilean satellites and give the possible origin of those features.

Study Activities

Key Terms and Concepts

Galilean satellites
liquid molecular hydrogen
liquid metallic hydrogen
magnetosphere
Great Red Spot
Io
Europa
Ganymede
Callisto

Using What You've Learned

1. Locate the planet Jupiter on the SC1 Star Chart. Use the longitudes given in the table in Lesson Eight. Use some date near the current one and locate the sun. Determine if the planet can be seen in the nighttime sky. Go out and observe it.

2. With a small telescope, observe Jupiter. Find the Great Red Spot and the banded structure of the atmosphere. Write down what you saw and compare it to the observations made by Galileo.

3. Based on observations made by the Pioneer and Voyager spacecraft, what conclusions were drawn about the nature of the Great Red Spot? You may wish to refer to the detailed descriptions given in magazine articles such as those listed in the bibliography in the back of this study guide.

4. With a pair of binoculars, observe the Galilean satellites. Identify their names and positions. The details of an activity of this kind can be found in the article "Observing Eclipses of Jupiter's Moons," *Sky and Telescope,* October 1976, p. 239.

Recommended Readings

For supplementary reading see the bibliography in the back of this study guide.

Self-Test

(Select the one best answer.)

Objective 1: Describe the size and shape of Jupiter.

1. The volume of Jupiter is approximately
   a. equal to that of the sun.
   b. 100 times larger than Earth's and 100 times smaller than the sun's.
   c. 1,000 times larger than Earth's and 1,000 times smaller than the sun's.
   d. 10 times larger than the sun's and 100 times larger than Earth's.

Objective 2: Describe the orbital and rotational characteristics of Jupiter.

2. The oblateness in the telescopic image of Jupiter is due to its
   a. chemical composition.
   b. rapid rotation.
   c. equatorial ring system.
   d. large number of satellites.

Objective 3: Describe the Red Spot.

3. A storm cloud in the upper atmosphere of Jupiter best describes
   a. the Red Spot.
   b. the colorful cloud belts.
   c. a shadow seen on the tops of the clouds.
   d. the rings of Jupiter.

Objective 4: Describe the banded structure of Jupiter's atmosphere.

4. Of the following reasons, which one do most scientists believe would prohibit life from evolving on Jupiter?
   a. no solid surface
   b. stronger gravity than Earth
   c. the planet's rapid rotation
   d. lack of water
   e. none of the above

Objective 5: Describe the magnetic field associated with Jupiter.

5. The magnetic field of Jupiter is
   a. about the same strength as Earth's.
   b. 20 times weaker than Earth's.
   c. 20 times stronger than Earth's.
   d. 1,000,000 times stronger than Earth's.

Objective 6: Describe the radiation associated with Jupiter.

6. Jupiter emits
   a. no radiation at all.
   b. more infrared energy than it receives from the sun.
   c. less infrared energy than it receives from the sun.
   d. gamma rays and X-rays.

Objective 7: Name the Galilean satellites and describe their characteristics.

7. Match the Galilean satellite with the characteristics listed.
   ____ 1. covered with impact craters
   ____ 2. active volcanos
   ____ 3. ice covered with numerous cracks
   ____ 4. largest satellite
   ____ 5. farthest from Jupiter
   ____ 6. affected by Jovian magnetic field and radiation belts

   a. Io
   b. Europa
   c. Ganymede
   d. Callisto

8. The satellite closest to Jupiter is
   a. Io.
   b. Europa.
   c. Amalthea.
   d. Callisto.

Objective 8: Describe the modern model of Jupiter's interior and atmosphere.

9. From the interior of Jupiter outward, the structure of the planet consists of
   a. liquid hydrogen, hydrogen and helium gas, metallic hydrogen, rocky surface.
   b. possible rocky core, metallic hydrogen, liquid hydrogen, hydrogen and helium gas.
   c. metallic hydrogen, rocky center, liquid hydrogen, hydrogen and helium gas.

10. The small obliquity of Jupiter (about 3 degrees) and its nearly circular orbit prevent Jupiter from having
    a. seasons.
    b. a magnetic field.
    c. life in its atmosphere.
    d. radiation belts.

Objective 9: Describe the rings of Jupiter.

11. Which of the following statements describe the ring system of Jupiter?
    a. has been seen from Earth
    b. more extensive than the rings of Saturn
    c. discovered by the Voyager spacecraft
    d. made up of tiny particles
    e. several hundred kilometers thick
    f. contains large-sized boulders

Answer Key

1. c
2. b
3. a
4. e
5. c
6. b
7. 1-c, d; 2-a; 3-b; 4-c; 5-d; 6-a
8. c
9. b
10. a
11. a, c, d

Saturn 11

Overview

"Among the myriad objects in the heavens," begins the program for this lesson, "one spectacular planet stands apart in its awesome beauty." Nearly as large as Jupiter, but much less massive, Saturn doesn't appear particularly special to the casual viewer. In the night sky, Saturn shines less brightly than Venus, Jupiter, some stars, and sometimes even Mars. Saturn does have the distinction of being the most distant planet visible to the unaided eye, so its journey along the zodiac (ecliptic) is most leisurely; Saturn revolves around the sun only once in almost thirty years.

Galileo's first views through his crude telescope revealed something of the structure that has fascinated astronomers and casual viewers ever since. As telescopes improved, more and more of the ring structure could be seen, and it soon was referred to as the "rings of Saturn." Three distinct rings, one inside the other, were identified from Earth-based observations, with a space or gap between the outer ring and the brightest, center ring. You've already learned how spectroscopy helped astronomers establish the general composition of the rings, which are made up of countless particles, each orbiting independently around the parent planet. Eventually, the gap between the outer and center rings was attributed to a phenomenon called resonance, an effect of the gravitational pull of Saturn's innermost large moon.

Saturn is host to a large number of satellites. For many years, only nine moons were known, but telescopic studies and, more recently, space probe explorations have identified more, with the total now at least nineteen. Still more satellites will probably be confirmed in the future. Titan, Saturn's largest moon, is the only satellite known to possess an atmosphere.

In this lesson, the combined arts of time-lapse photography, computer processing, and animation make it possible for you to retrace Voyager's journey past Saturn and its moons, and even cross its ring plane. You'll learn that, though Saturn might be thought of as a smaller brother to Jupiter — Saturn also has huge cloud bands and "Red Spot" eddy systems — it sports winds with much greater velocity than those of Jupiter, for example. You'll also study some of the discoveries made from photographs at very close range of Titan and Saturn's other larger moons. One of the most intriguing discoveries of the Voyager missions was the complexity of the ring structure. Not only were new rings discovered, but the large visible ring was found to be made up of many concentric "ringlets," looking something like a grooved phonograph record.

In the television program you'll take a close look at the Voyager spacecraft. You'll see how its major systems are attached and where many of its important experiments were placed. As in the program on Jupiter, there will also be descriptions of some of the nonvisual experiments performed by the spacecraft at Saturn. Maybe you'll find it worth a moment's reflection that space-age astronomy has somehow made the planets more fascinating, even while it has posed some more difficult questions for astronomers.

Learning Objectives

When you have completed all assignments in this lesson, you should be able to:

1. Describe the size and shape of Saturn.

2. Describe the orbital and rotational characteristics of Saturn.

3. Describe Saturn's ring system, name the different rings, and know the two possible origins of the rings.

4. Describe the banded structure of Saturn's atmosphere.

5. Describe the characteristics of Saturn's largest satellite, Titan.

6. Describe the most important discoveries of the Voyager missions to Saturn.

7. Describe the objectives and successes of the Pioneer missions to Saturn.

Assignments for This Lesson

Before Viewing the Program

Read the overview at the beginning of this lesson and look over the learning objectives that precede the assignments.

Read text Chapter 10, pages 187-197 and 199-202, and pages 203-205, 210-213, and 214-219 in Chapter 11 of the text.

View Program 11, "Saturn"

As you view the program, look for:

- how we gained our first understanding of the nature of Saturn's rings.

- images of the rings taken by Voyager spacecraft, specifically, the ringlets, spokes in the B ring, and the braided F ring.

- images of Saturn's moons taken by the Voyager spacecraft and what these photographs have told us about the nature of these moons.

- new information on the nature of Titan's atmosphere and surface.

- images of the cloud tops of Saturn taken by the Voyager spacecraft.

After Viewing the Program

Review the text assignment.

Review the terms that appear in italics and boldfaced type in the text. In particular, note the key terms and concepts listed in the study activities section.

Complete the study activities and the self-test in this study guide; then check your answers with the answer key at the end of the lesson.

Complete the following exercises:

1. Explain how Saturn compares in size with the other eight planets. In particular, how much larger is Saturn than Earth and how much smaller is Saturn than Jupiter? Explain why Saturn is so oblate.

2. Cite the approximate periods of Saturn's revolution in years and rotation in hours. You should also be able to compare these with the revolutions and rotations, respectively, of the other planets.

3. Cite the approximate inner and outer diameters of Saturn's rings, describe their composition, and identify the A, B, and C rings. Locate Cassini's division. You should also understand and be able to explain the two theories regarding the origin of the rings.

4. Cite the approximate size of Titan, relative to the other satellites and planets in the solar system, and describe its composition. Describe the composition and approximate density of Titan's atmosphere.

5. Complete exercises 4 and 6 on page 221 in text Chapter 11.

Study Activities

Key Terms and Concepts

Cassini's division
Roche limit
Saturn's rings
Titan
ringlets
spokes
shepherding satellites
F ring
radar astronomy

Using What You've Learned

1. Using a current astronomy magazine to locate its position, go out at night and try to observe Saturn. Good binoculars or a small telescope should also permit the observation of Saturn's rings and its large moon, Titan. Make a sketch of your observations. Try to include Cassini's division in your sketch. (You may also try to use the coordinates in study guide Table 8.1 and the SC1 Star Chart.)

2. With the aid of a small telescope of at least 7 cm aperture, or very good binoculars, observe the large moon, Titan, over a period of a month. Initially, it may be difficult to distinguish Titan from the background stars that appear near Saturn. However, by sketching Saturn on successive nights and locating the nearby stars as seen in the telescope, and then comparing the sketches, you will observe two objects — Saturn and Titan — appearing to move together relative to the stationary background stars. Trace their paths relative to the background stars and relative to each other. From this, you should be able to estimate the period of Titan's revolution around Saturn. How does this period compare with those of the Galilean satellites and why is Titan's period different?

Recommended Readings

For supplementary reading see the bibliography in the back of this study guide.

Self-Test

(Select the one best answer.)

Objective 1: Describe the size and shape of Saturn.

1. How does Saturn's size, minus the rings, compare with the size of Earth and Jupiter?
   a. Saturn is smaller than both.
   b. Saturn is smaller than Jupiter but about equal to Earth.
   c. Saturn is slightly smaller than Jupiter and about 9.5 times larger than Earth.
   d. Saturn is about equal to Jupiter in size, and therefore about 11 times larger than Earth.
   e. Saturn is larger than both Jupiter and Earth.

2. Which is true about the density of Saturn?
   a. Like the other Jovian planets, it is less dense than water.
   b. It is the only planet that is less dense than water.
   c. It is about twice as dense as water.
   d. Like the terrestrial planets, it is about five times more dense than water.
   e. Saturn's density is unknown.

3. Which of the descriptions given below are *not* characteristic of the planet Saturn?
   a. It is made mostly of hydrogen and helium.
   b. It has an extensive ring system.
   c. It has the most number of known *large* moons.
   d. It is the farthest from the sun of the six planets easily visible to the naked eye on Earth.
   e. It is the brightest planet in the sky, as seen from Earth.
   f. It is the fastest rotating planet.
   g. It has a banded atmospheric structure.

Objective 2: Describe the orbital and rotational characteristics of Saturn.

4. What feature in the telescopic appearance of Saturn indicates it is a relatively fast rotator?
   a. the banded structure of the atmosphere
   b. its color
   c. the fact that it has rings
   d. its oblateness
   e. the large number of satellites
   f. the large size of the planet

5. What are the periods of Saturn's rotation and revolution, respectively?
   a. Both are about 10 years.
   b. Its rotation period is about equal to Jupiter's, but its revolution period is twice that of Jupiter.
   c. Its rotation period is unknown, but its revolution period is about 29.5 years.
   d. They are about 10.23 hours and 29.5 years, respectively.
   e. They are about 29.5 days and 10.23 years, respectively.

Objective 3: Describe Saturn's ring system, name the different rings, and know the two possible origins of the rings.

6. Which three of the following are observations which show that Saturn's rings do not rotate as one solid object?
   a. bright stars can be observed to shine through the rings
   b. the fact that Saturn has at least 19 moons
   c. differential rotation of the ring features observed by spacecraft
   d. Saturn's rapid rotation
   e. the fact that Saturn emits more energy than it receives from the sun
   f. Doppler shift in the spectra from the rings shows that the inner edge of the ring moves faster than the outer edge

7. How might Saturn's rings appear to an observer who is "standing" on the planet's equator at midnight? (Saturn has no surface to stand on, of course.)
   a. a bright, wide band across the sky
   b. a dark, wide band across the sky
   c. a bright line across the sky from horizon to horizon
   d. two straight, thin bright lines, extending up from the E and W points on the horizon, toward the zenith
   e. a dark line across the sky, from horizon to horizon
   f. They would be invisible.

8. Which two of the following are accepted as possible origins of Saturn's rings?
   a. a collection of comet remnants, attracted by Saturn's strong gravity
   b. a series of communication satellites put up by the Saturnians
   c. a moon which came too close to Saturn and was pulled apart by the planet's gravity
   d. remnants of an exploded satellite of Saturn
   e. the remains of proto-Saturn which just never formed into one or more large bodies
   f. Astronomers have no idea from where the rings might have come.

Objective 4: Describe the banded structure of Saturn's atmosphere.

9. Which of the reasons listed below make Saturn an unpleasant place for humans to live?
   a. too cold
   b. too hot
   c. no atmosphere to breathe
   d. no oxygen to breathe
   e. no liquid water to drink
   f. no solid surface to stand on
   g. surface gravity is too strong

Objective 5: Describe the characteristics of Saturn's largest moon, Titan?

10. Which of the following describe Saturn's largest moon, Titan?
   a. It is larger than Mercury.
   b. It is easily visible to the naked eye.
   c. It is known to possess a dense atmosphere.
   d. Its opaque atmosphere hides its surface features.
   e. It has an unusual, polar orbit around Saturn.
   f. It is the second largest of Saturn's satellites.

Objective 6: Describe the most important characteristics of the Voyager missions to Saturn.

11. Which of the following was *not* discovered by the Voyager 1 encounter with Saturn?
    a. several new satellites
    b. Cassini's division is indeed clear of material
    c. the large rings are actually hundreds of thin rings
    d. many of the satellites are extensively cratered
    e. radial, spoke-like features in the rings

12. Which were *not* observed or discovered by the Voyager 2 encounter with Saturn?
    a. the satellite Phoebe is round
    b. the braided F ring observed by Voyager 1 had become a single strand
    c. craters on the surface of Titan
    d. large shepherding satellites within the rings to account for the detailed ring structure
    e. the large rings are actually thousands of thin rings

Objective 7: Describe the objectives and successes of the Pioneer missions to Saturn.

13. What were two main goals of the Pioneer 11 mission to Saturn?
    a. a close examination of Titan
    b. a close examination of the region between Saturn and the rings
    c. a study of Saturn, its rings, and the environment around the planet
    d. a preliminary scouting for the more sophisticated Voyager missions
    e. a search for life on Saturn
    f. a search for life on the rings

14. Why did the Pioneer 11 and Voyager spacecraft fly by Jupiter en route to Saturn?
    a. There's no special reason; Jupiter just happens to lie between Earth and Saturn.
    b. The spacecraft use Jupiter's magnetic field to propel themselves on to Saturn.
    c. The spacecraft use Jupiter's gravitational field to help propel themselves on to Saturn.
    d. Jupiter's strong radiation field assists the spacecraft on to Saturn.

Answer Key

1. c
2. b
3. c, e, f
4. d
5. d
6. a, c, f
7. d
8. c, e
9. a, d, e, f
10. a, c, d
11. b
12. c, d
13. c, d
14. c

Uranus, Neptune, and Pluto 12

Overview

With this lesson you enter that part of the realm of the planets most recently to be explored by twentieth-century spacecraft. Uranus was explored by Voyager 2 in January 1986. If all goes well, Voyager 2 will be sending photographs and data from the last of the Jovian planets during its encounter with Neptune in August 1989. If past experience is any guide, a successful flyby of Neptune will answer many of the questions astronomers have today, and lead to new questions about the formation of this planet and the entire solar system.

This lesson reaches out to one boundary of the solar system, the orbit of Pluto. This seems a good time to consider again the immense distances involved in our solar system. Remember the speed of light? Travelling at about 186,200 miles per second, light from the sun reaches Earth in about eight minutes. It takes about five and one-half hours for light to reach Pluto at its average distance from the sun. You may find the speed of light a bit difficult to comprehend. If so, let's consider a more reasonable speed, the escape velocity for Earth of about 25,000 miles per hour (the speed at which Apollo astronauts received their boost out of Earth's orbit toward the moon). Imagine an Apollo spacecraft launched from the sun at 25,000 miles per hour, moving directly out from the sun toward Earth (we violate some practical requirements of space travel, of course, to achieve such straight-line velocity). This imaginary ship would have to travel 161 days or over five months to reach Earth. Now, imagine this craft launched at the same speed toward Pluto; its travel time will have to be logged in years, rather than days or months. At 25,000 mph, travel time from the sun to Pluto's mean distance is slightly over seventeen and one-half years.

If you're fortunate enough to find really good viewing conditions, a clear dark sky away from city lights, and you know where to look, you might be able to see Uranus as a starlike point of light (just at the limits of visibility). Over a period of several nights, you'd observe this one point of light moving slowly in relation to the background stars, which would confirm that you were indeed looking at Uranus, since Neptune and Pluto can be seen only through a telescope.

This lesson attempts to illustrate that careful observation is the key to astronomical discovery. You'll learn that Uranus was discovered by accident; Neptune was discovered through careful calculations based on irregularities in Uranus's orbit; and Pluto's discovery, though it was also based on careful calculations, turned out to be a fortunate coincidence. You'll learn how unexpected phenomena, far too minute to be observed casually through the telescope, can still be detected by the alert astronomer. Such unexpected findings include not only Uranus, but Uranus's rings and Pluto's moon.

Our knowledge of Uranus has increased dramatically since the Voyager flyby, as you'll read in your text. Although Neptune has only been studied through telescopes, a surprisingly large amount of information has also been acquired about this distant world. Tiny Pluto doesn't fit neatly into the pattern set by the other eight planets; you'll learn about a hypothesis, yet to be accepted as a satisfactory model, indicating that Pluto may once have been a satellite of Neptune's. Answers are still sought for questions about these planets. At the same time, the search goes on for yet another planet beyond Pluto's distant orbit.

Learning Objectives

When you have completed all assignments in this lesson, you should be able to:

1. List and relate the historical events which led to the discoveries and namings of Uranus, Neptune, and Pluto.

2. Describe the physical properties and satellite systems of Uranus, Neptune, and Pluto, and compare them with each other.

3. Describe the orbital and rotational characteristics of Uranus, Neptune, and Pluto.

4. Describe some of the searches for additional planets in the solar system subsequent to Pluto's discovery.

5. Describe the discovery of the ring system of Uranus.

6. Describe the discovery of the moon around Pluto.

7. Identify and describe the discoveries about Uranus made during the Voyager flyby in 1986.

8. Describe the surface features and characteristics of the moons of Uranus as discovered by the Voyager spacecraft.

Assignments for This Lesson

Before Viewing the Program

Read the overview at the beginning of this lesson and look over the learning objectives that precede the assignments.

Read carefully "Discovery of Neptune" in text Chapter 2, page 44.

Read the sections on Uranus, Neptune, and Pluto, pages 197-202 and 210-221 in Chapters 10 and 11 of the text.

View Program 12, "Uranus, Neptune, and Pluto"

As you view the program, look for:

- the explanation of how each of these planets was discovered.

- the description of the discovery of the rings of Uranus.

- the orientation of the axis of rotation and the orbital plane of the moons of Uranus.

- the description of the peculiar orbits of the moons of Neptune.

- the description of the unusual characteristics of the orbit of Pluto and the discovery of a moon orbiting the planet.

- the search being conducted for planets beyond Pluto and the discovery of Charon.

After Viewing the Program

Review the text assignments.

Review the terms that appear in italics and boldfaced type in the text. In particular, note the key terms and concepts listed in the study activities section.

Complete the study activities and the self-test in this study guide; then check your answers with the answer key at the end of the lesson.

Complete the following exercises:

1. Describe in detail the following peculiarities of the three outer planets:
   a. Uranus — tilt in its rotational axis;
   b. Neptune — moons Triton and Nereid;
   c. Pluto — inclination and eccentricity of its orbit around the sun.

2. Match the following planets with the correct descriptions.
   Which planet:

_____ 1. has the largest number of known satellites?	a. Jupiter
_____ 2. has the famous red spot?	b. Saturn
_____ 3. has the surface with the highest temperature?	c. Venus
_____ 4. has frozen water and evidence of former rivers?	d. Mars
_____ 5. has the smallest mass?	e. Pluto
_____ 6. has the largest mass?	f. Earth
_____ 7. is farthest from Earth?	g. Mercury
_____ 8. has polar caps?	h. Uranus
_____ 9. has the strongest magnetic field?	i. Neptune
_____ 10. has a satellite with active volcanos?	
_____ 11. is third from the sun?	
_____ 12. has rings?	
_____ 13. is most like Earth in size?	
_____ 14. is most like Earth in surface conditions?	
_____ 15. is closest to the sun?	
_____ 16. was most recently discovered?	
_____ 17. was discovered by mathematical prediction?	
_____ 18. is associated with Cassini's division?	
_____ 19. has the largest amount of carbon dioxide in its atmosphere?	
_____ 20. has the lowest mean density?	

Study Activities

Key Terms and Concepts

perturbation
occultation
Charon

Using What You've Learned

1. Describe the seasons as they would occur on Uranus and compare them to Earth's.

2. Outline the events which led to the discoveries of Uranus, Neptune, and Pluto.

3. Explain why Pluto, for the time being, is not the most distant planet in the solar system.

Recommended Readings

For supplementary reading see the bibliography in the back of this study guide.

Self-Test

(Select the one best answer.)

Objective 1: List and relate the historical events which led to the discoveries and namings of Uranus, Neptune, and Pluto.

1. The discovery of Pluto was hindered by the fact that
   a. it is small and very faint.
   b. changes in its position are small and therefore not easily detected.
   c. it has little gravitational influence over the other members in the solar system.
   d. all of the above are true.

2. The discovery of the planet ____________ was made by accident during a survey of the sky by William Herschel.
   a. Neptune
   b. Uranus
   c. Pluto

3. Analysis of the gravitational influences on Uranus led to the discovery of
   a. Neptune.
   b. Pluto.
   c. Pluto's moon.
   d. Triton and Nereid.

4. Pluto was discovered during a planetary search by
   a. William Herschel.
   b. J.C. Adams.
   c. Clyde Tombaugh.
   d. Charles Kowal.

Objective 2: Describe the physical properties and satellite systems of Uranus, Neptune, and Pluto, and compare them with each other.

5. Which of the following describes the planet Uranus?
   a. thin ring system
   b. accidental discovery
   c. equator of the planet lies nearly perpendicular to the plane of its orbit
   d. Voyager flyby in 1986
   e. all of the above

6. Which of the following describes the planet Neptune?
   a. position predicted mathematically
   b. largest planet in the solar system
   c. moves farthest from the sun
   d. has one large moon

7. Which of the following describes the planet Pluto?
   a. the smallest planet in the solar system
   b. has one moon
   c. moves farthest from the sun
   d. cold enough for atmospheric gases to lie frozen on it surface
   e. all of the above

8. The atmospheres of Uranus and Neptune consist of
   a. hydrogen, chlorine, methane, carbon.
   b. hydrogen, helium, ammonia, methane.
   c. argon, chlorine, methane, ammonia.
   d. carbon dioxide, neon, hydrogen, oxygen.

9. Of what do the interiors of Uranus and Neptune consist?
   a. similar to the interior of Earth
   b. solid rock throughout
   c. liquid hydrogen throughout
   d. core rocky, surrounded by layers of ice, surrounded by layers of liquid hydrogen

Objective 3: Describe the orbital and rotational characteristics of Uranus, Neptune, and Pluto.

10. Of the three outer planets, which one has the orbit with the highest inclination and largest eccentricity?
    a. Neptune
    b. Uranus
    c. Pluto

Objective 4: Describe some of the searches for additional planets in the solar system subsequent to Pluto's discovery.

11. During a systematic search of the outer solar system, Charles Kowal has discovered another planet beyond the orbit of Pluto.
    a. true
    b. false

Objective 5: Describe the discovery of the ring system of Uranus.

12. The rings of Uranus were discovered during
    a. a search for comets.
    b. a search for additional moons around the planet.
    c. the observation of an occultation of a star by Uranus.
    d. a search for the predicted rings around the planet.

Objective 6: Describe the discovery of the moon around Pluto.

13. How was the moon orbiting Pluto discovered?
    a. visually
    b. photometrically
    c. spectroscopically
    d. photographically

Objective 7: Identify and describe the discoveries about Uranus made during the Voyager flyby in 1986.

14. From Earth-based observations, nine thin rings were discovered around Uranus. The Voyager spacecraft discovered an additional ______________ rings.
    a. 1
    b. 5
    c. 2
    d. 0

15. The structure of the Epsilon Ring about Uranus is explained by the existence of
    a. gravity
    b. Miranda
    c. dust particles in the ring
    d. shepherd satellites

Objective 8: Describe the surface features and characteristics of the moons of Uranus as discovered by the Voyager spacecraft.

16. Great valley systems with gorges 10 km deep and trapezoidal mountain ranges describes most nearly the moon
    a. Miranda
    b. Umbrial
    c. Airel
    d. Oberon

Answer Key

1. d
2. b
3. a
4. c
5. e
6. a
7. e
8. b
9. d
10. c
11. b
12. c
13. d
14. c
15. d
16. a

Solar System Debris 13

Overview

In Lesson Ten it was suggested that an astronomer from somewhere else in the universe might consider Jupiter to be the sun's only important companion, and all other bodies "debris" from the solar system's formation. Even from our viewpoint, there's a tendency to think of the smaller objects in the solar system as "leftovers." These objects range in size from that of a very small moon down to dust particles. Some of the objects are responsible for such startling visual effects as those caused by comets, meteors, and the zodiacal light. Others are invisible to the unaided eye and could be overlooked even if photographed through a telescope. Why does all this debris attract so much attention? There aren't any definite answers yet concerning its origins, and, furthermore, these objects and particles contain more clues on how the solar system itself was formed.

The first class of objects you'll study in this lesson are the asteroids, those irregularly shaped objects that orbit the sun primarily between the orbits of Jupiter and Mars in the "asteroid belt." Other asteroids precede and follow Jupiter in its orbit (the Trojan asteroids); still others follow elongated orbits that take them across Earth's orbit (the Apollo asteroids) or, in some instances, even the orbit of Venus. Astronomers have gathered evidence suggesting that these bodies were once part of a much larger planet; however, there is also evidence suggesting that these bodies consist of materials indicating they were never part of larger bodies. And it is possible that all asteroids do not have a common origin.

You'll also be introduced to the comets, those "hairy stars" that have created so much superstitious and (in recent years) pseudoscientific fear. On rare occasions, the glowing head and tail of a comet can stretch across much of the night sky, but most are much less impressive, if they can be seen at all. You will read about the space missions which flew by Halley's comet in 1986. These missions were planned by a European consortium, Japan, and the Soviet Union. The United States did not send a spacecraft to Halley. The glowing head and tail of a comet are made up of vaporized gases and freed particles which leave the comet nucleus as it heats up from exposure to sunlight. The nucleus has been described as a "dirty snowball" or "iceberg," which gives you some idea of its composition.

The remaining class of debris is related both to asteroids and comets. Most of the larger meteoroids are believed to be chips off asteroids. When the meteoroid reaches Earth's atmosphere, it becomes hot and glows because of air friction. The resulting flash of light is called a meteor. If the chip or particle is large enough to survive the flight through the atmosphere and is found on the surface of Earth, it is called a meteorite. Comets also leave particles as small as grains of sand in space. Should Earth encounter this debris, the particles will become hot and glow. When many glowing particles such as these are seen over a short period of time, an Earth-based observer is treated to a meteor shower.

In the television program, you'll examine specimens of the three types of meteorites and learn about the origins of each of these types. You should appreciate the importance of meteorites to the science of astronomy. After all, until the Apollo missions brought back some rock specimens from the moon (and Viking tested some soil samples on Mars), the meteorites were the only extraterrestrial materials available for laboratory testing and experiments.

In this lesson you'll also have a quick look at some optical aids that astronomers have developed to help them in their continuing search of space.

Learning Objectives

When you have completed all assignments in this lesson, you should be able to:

1. Describe the events leading to the discovery of the first asteroid.
2. Describe the properties, characteristics, classifications, and locations of the asteroids.
3. Cite the possible origins of asteroids.
4. Describe the properties and origins of meteorites.
5. Name and describe the four basic parts of a comet.
6. Describe how the solar wind interacts with the tail of a comet.
7. Describe the characteristics of a meteor and a meteor shower.
8. Describe the characteristics of cometary orbits and their relationship with meteor showers.
9. Describe the nature and significance of the Oort cloud.
10. Identify and describe the discoveries made by the spacecraft flybys of Halley's comet apparition of 1986.

Assignments for This Lesson

Before Viewing the Program

Read the overview at the beginning of this lesson and look over the learning objectives that precede the assignments.

Read all of text Chapter 12 and pages 241-246 of Chapter 13.

View Program 13, "Solar System Debris"

As you view the program, look for:

- the description of the objects known as minor planets (asteriods) located between the orbits of Mars and Jupiter. Be sure to note the two theories of the formation of the asteroids.

- the description of the different types of meteorites and what kinds of information these objects reveal.
- the description of the historical importance of comets.
- the description of the various parts of a comet (tail, nucleus, head) and the cause of a comet's tail.
- the explanation of the formation of meteor showers.
- the theories of the origin of comets and the Oort cloud.
- final comments about the effects comets may have on Earth and the death of the dinosaurs.

After Viewing the Program

Review the text assignment.

Review the terms that appear in italics and boldfaced type in the text. In particular, note the key terms and concepts listed in the study activities section.

Complete the study activities and the self-test in this study guide; then check your answers with the answer key at the end of the lesson.

Complete the following exercises: 3, 7, and 9 on page 240 in Chapter 12 of the text.

Study Activities

Key Terms and Concepts

asteroid (minor planet, planetoid)
Apollo asteroids
asteroid belt
Trojan asteroids
comet
Oort cloud
nucleus (of a comet)
tail (of a comet)
coma (of a comet)
radiation pressure
meteor
meteorite
meteoroid
shower meteors (meteor showers)
fireballs (bolides)
zodiacal light

Using What You've Learned

1. Observe a meteor shower. Use Table 12.1 in the text on page 239 to select the shower best suited to your needs. Keep track of the number of meteors per hour. Determine the radiant point of the shower.

2. Will Halley's comet last indefinitely? Why or why not? If not, what might be expected to continue after the comet's time is over?

Recommended Readings

For supplementary reading see the bibliography in the back of this study guide.

Self-Test

(Select the one best answer.)

Objective 1: Describe the events leading to the discovery of the first asteroid.

1. The asteroid Ceres was discovered
   a. accidentally, during a search for comets.
   b. photographically.
   c. on Earth as a meteorite.
   d. during a search for the "missing planet."

Objective 2: Describe the properties, characteristics, classification, and locations of the asteroids.

2. Most asteroids are located between the orbits of
   a. Jupiter and Saturn.
   b. Mars and Jupiter.
   c. Saturn and Uranus.
   d. Earth and Venus.

3. Which best describes the shape of large asteroids such as Ceres and Vesta?
   a. oblate
   b. long and thin
   c. irregular
   d. round

Objective 3: Cite the possible origins of asteroids.

4. The formation of asteroids is the result of
   a. the collision between two planets.
   b. planetary material that never formed into large planets.
   c. an exploded planet.
   d. the escape of Jovian type satellites.

Objective 4: Describe the properties and origins of meteorites.

5. Pieces of asteroids that land on Earth are called
   a. meteors.
   b. meteoroids.
   c. meteorites.
   d. tektites.

6. Which type of meteorite has the highest density?
   a. irons
   b. stones
   c. stony irons

7. The scarcity of impact craters on Earth is due to
   a. the low number of actual impacts.
   b. shielding by the moon.
   c. erosion and geological processes destroying craters that were present in the past.
   d. the fact that most impacts occurred in the ocean, leaving no craters.

Objective 5: Name and describe the four basic parts of a comet.

8. Of the following, which is *not* a part of a comet?
   a. coma
   b. nucleus
   c. solar wind
   d. tail
   e. hydrogen cloud

Objective 6: Describe how the solar wind interacts with the tail of a comet.

9. The interaction of the solar wind and the gases of a comet produce
   a. a tail that points away from the sun at all times.
   b. a tail that points toward the sun at all times.
   c. the coma.
   d. nothing.

Objective 7: Describe the characteristics of a meteor and a meteor shower.

10. When Earth encounters random meteoroids, the result is
    a. a meteor shower.
    b. sporadic meteors.
    c. a cometary display.
    d. satellite occultations.

11. What is the result of Earth encountering the remnant of a comet?
    a. meteorites
    b. impact craters
    c. production of Widmanstatten lines
    d. meteor showers

Objective 8: Describe the characteristics of cometary orbits and their relationship with meteor showers.

12. A comet's orbit
    a. is nearly circular.
    b. lies between Saturn and Uranus.
    c. is highly elliptical, cigar-shaped.
    d. is geometrical.

Objective 9: Describe the nature and significance of the Oort cloud.

13. Most scientists feel that comets originate
    a. around other stars.
    b. from what is known as the Oort cloud.
    c. in the vicinity of Uranus.
    d. in the vicinity of Pluto.

Objective 10: Identify and describe the discoveries made by the spacecraft flybys of Halley's comet apparition in 1986.

14. The Vega and Giotto spacecraft observed which of the following on Comet Halley in 1986?
    a. living organisms
    b. liquid water
    c. "jets" resembling volcanic plumes and geysers
    d. a very dark and irregularly shaped nucleus

15. Much of the dust detected by the Giotto spacecraft at closest approach to the nucleus of Comet Halley consisted of
    a. carbon and hydrocarbon compounds
    b. silicates
    c. gold and silver
    d. iron and nickel

Answer Key

1. d
2. b
3. d
4. b
5. c
6. a
7. c
8. c
9. a
10. b
11. d
12. c
13. b
14. c, d
15. a

The Solar Image 14

Overview

That two lessons are devoted to a single object — the sun — is a natural reflection of the sun's complexity and an indication of its paramount importance to Earth and of the influence it has over our daily lives. We take the light and heat from the sun for granted (unless they interfere with our activities). Yet the sun affects processes in much more subtle ways — photosnythesis in plants is the most obvious example, but not the only one. Particles streaming from the sun (the "solar wind") can affect radio communications and even electronic instruments.

In this lesson you'll learn about a seventy-year period during which the sun appears to have had virtually no sunspots. A year or so of a "quiet sun" is a normal part of the sunspot cycle which astronomers have followed for many years, but such a long period of quiet is unusual. This quiet period coincided with an extended period of colder weather on parts of Earth, a time that some meteorologists have dubbed "the mini-ice age." It's not yet clear what connection exists between the two, but it raises the suspicion that life on Earth is dependent on the sun in ways that have not yet been discovered.

There's yet another reason for studying the sun: its proximity to us. Light from the sun reaches Earth in about eight minutes. This short amount of time makes it easier for astronomers to study the sun in great detail. The sun is our nearest star and our next nearest stellar neighbor is considerably farther away. Light from that star takes about four years to arrive at Earth. Even when it arrives, it can be studied only as a point of light. The techniques of speckle interferometry (discussed in a later lesson) and orbiting observatories may make it possible for astrophotography to resolve an image of some very large, not-too-distant stars into a small disk, but no other star is going to reveal to us the detail that can be seen on our sun.

In this lesson you will learn the dimensions of our sun and its period of rotation. You'll be concentrating on the "visible" portion of the sun, its atmosphere. However, you'll learn that much of the visible part of the sun can't be seen except under very special circumstances. The outer regions of the sun's atmosphere are tenuous and transparent and can be seen only during a total solar eclipse. There is much to studied at the "surface" of the sun (the lowest portion of the solar atmosphere, or photosphere). Besides the sunspots, there are granulations, supergranules, solar flares, and prominences on the photosphere. Each of these phenomena provides clues to activities beneath the sun's surface.

Even though the sun is the brightest object in our sky, the atmosphere surrounding Earth makes it difficult to see the sun in great detail. A small portion of the television program shows two ways astronomy overcomes this difficulty. One obvious way is to make observations from outside the atmosphere; and that, of course, is what the orbiting solar observatories do. A more earthbound alternative is to find an area on Earth that does not experience great changes in temperature during daylight hours; this solution accounts for the unusual location of the Big Bear Solar Observatory in the Southern California mountains. Be sure to watch for the discussion of this observatory in the television program.

Learning Objectives

When you have completed all assignments in this lesson, you should be able to:

1. Describe the size, shape, and rotation of the sun.

2. Name and describe the major regions of the sun and their temperature variations.

3. Describe photospheric granulations, sunspots, and the sunspot cycle.

4. Describe solar surface flares and prominences.

5. Describe the nature of the solar wind.

6. Locate and understand the nature of the photosphere.

7. Describe the nature of coronal holes.

Assignments for This Lesson

Before Viewing the Program

Read the overview at the beginning of this lesson and look over the learning objectives that precede the assignments.

Read Chapter 17, pages 298-310 and 311-315, in the text.

View Program 14, "The Solar Image"

As you view the program, look for:

- a description of the instruments used to study the sun.
- the explanation of the terms used to describe the atmosphere of the sun (photosphere, chromosphere, corona).
- the description and explanation of the various features seen on the photosphere, including prominences, convection granules, sunspots, and flares.
- the relationship between coronal holes and the solar wind.
- the sunspot cycle and the Maunder minimum.
- the effects that solar flares have on Earth (auroras, communication interference).

After Viewing the Program

Review the text assignment.

Review the terms that appear in italics and boldfaced type in the text. In particular, note the key terms and concepts listed in the study activities section.

Complete the study activities and the self-test in this study guide; then check your answers with the answer key at the end of the lesson.

Complete the following exercises: 2, 4, 6, and 7 in text Chapter 17, page 315.

Study Activities

Key Terms and Concepts

photosphere
flash spectrum
chromosphere
spicules
corona
sunspots
granulation
sunspot cycle
differential rotation
Maunder minimum
prominences
flares
solar wind
coronal holes
aurora
convection
solar constant
spectroheliograph

Using What You've Learned

1. Observe the sun through a small telescope. This can be done by projecting the image of the sun onto a white sheet of paper. DO NOT LOOK THROUGH THE TELESCOPE AT OR NEAR THE SUN. Sketch the sun, making note of the position of sunspots. Do this at the same time for several days. Try to determine the rate of rotation of the sun.

2. Investigate the various effects the Maunder minimum may have had on Earth.

3. Describe the various instruments and satellites that have been and are being used to study the sun.

4. Describe the nature of coronal holes and the part they play in the production of the solar wind.

Recommended Readings

For supplementary reading see the bibliography in the back of this study guide.

Self-Test

(Select the one best answer.)

Objective 1: Describe the size, shape, and rotation of the sun.

1. The sun is
   a. about 10 times larger than Jupiter, with no measurable oblateness.
   b. about the same size as Jupiter, but twice as oblate.
   c. about 100 times larger than Jupiter, with the oblateness of Saturn.
   d. about 100 times larger than Jupiter, with no measurable oblateness.

2. The rotation of the sun is
   a. uniform throughout.
   b. faster at the poles than at the equator.
   c. faster at the equator than at the poles.
   d. equal to the rotation rate of Earth.

Objective 2: Name and describe the major regions of the sun and their temperature variations.

3. Dark regions on the surface of the sun are called
   a. granules.
   b. sunspots.
   c. spicules.
   d. plages.

Objective 3: Describe the photospheric granulations, sunspots, and the sunspot cycle.

4. The sunspot cycle refers to the
   a. length of time for the sun to turn on its axis once.
   b. movement of sunspots on the solar disk.
   c. length of time for the average number of sunspots to vary from a maximum to minimum and back to maximum.
   d. time of appearance of sunspots on the solar disk.

5. The layer of the sun that is normally seen is the
   a. chromosphere.
   b. corona.
   c. solar wind.
   d. photosphere.

6. The tops of convection columns of hot, rising gas appear in the photosphere as
   a. granules.
   b. flares.
   c. sunspots.
   d. spicules.

Objective 4: Describe solar surface flares and prominences.

7. Small regions on the sun where huge amounts of energy are suddenly released are called
   a. sunspots.
   b. granules.
   c. filaments.
   d. flares.

8. The large "flames" seen on the edge of the sun's disk are
   a. spicules.
   b. prominences.
   c. flares.
   d. granules.

9. The sun's outer atmosphere seen most easily during a total solar eclipse is the
   a. corona.
   b. chromosphere.
   c. photosphere.
   d. solar wind.

10. The atmosphere of the sun studied by what is known as the flash spectrum is the
   a. corona.
   b. chromosphere.
   c. photosphere.
   d. solar wind.

Objective 5: Describe the nature of the solar wind.

11. Charged particles leaving the sun are called
   a the solar wind.
   b. the corona.
   c. prominences.
   d. flares.

Objective 6: Locate and understand the nature of the photosphere.

12. When viewing the photosphere through a hydrogen alpha filter, a pattern of light and dark areas (similar to a checkerboard) is observed. These areas are known as
    a. solar flares.
    b. spicules.
    c. sunspots.
    d. granules.

Objective 7: Describe the nature of coronal holes.

13. In an X-ray photograph of the sun, coronal holes would appear as
    a. bright spots.
    b. small dark spots similar to sunspots.
    c. large dark regions.
    d. bands or belts in the atmosphere of the sun.

14. Coronal holes are believed to be the source of the
    a. neutrino flux.
    b. solar wind.
    c. visible light emitted by the sun.
    d. X-rays emitted by the sun.

Answer Key

1. a	8. b
2. c	9. a
3. b	10. b
4. c	11. a
5. d	12. d
6. a	13. c
7. d	14. b

The Solar Interior 15

Overview

The physical laws governing the sun's production of energy govern the life of the other stars as well, and it is believed that the processes which take place at the sun's core are duplicated in the "main sequence" stars you'll study in later lessons. Even the study of the model which describes the formation of the solar system is, in a way, an introduction to the stars, because the model, if it is correct, should explain how stars are formed throughout the universe.

An understanding of the processes occurring within our sun had to await the development of atomic theory and Einstein's theory of relativity. According to Einstein's theory, in the equation $E = mc^2$, a small amount of matter could be converted into a tremendous amount of energy. In the equation, "E" stands for energy and "m" stands for mass, which is multiplied by "c^2," the speed of light squared. The beauty of this theory, and the currently accepted model of the sun's energy production in its interior, is that it accounts both for the tremendous amount of energy produced and for the extremely long life span of the sun.

In this lesson you'll learn that the model of the formation of the solar system is based on currently accepted theories which have passed the tests of extensive observation. The theories, for example, explain the characteristics and properties of most objects within the solar system.

In this lesson you will also take a much more intensive look at the core of the sun, which was mentioned briefly in the preceding lesson. In fact, to understand what's happening there, you will have to consider the atoms of the sun. You'll recall that the sun is composed predominantly of hydrogen, with most of the rest composed of helium. In actuality, deep in the sun's interior, hydrogen atoms continuously combine to form helium atoms. Thus, eventually the hydrogen in the sun must become depleted. (This event will be given consideration in Lesson Twenty-four.) You'll also learn how energy from the sun's core is physically transferred out to and beyond the photosphere.

Still another part of this lesson recalls your study of electromagnetic radiation and spectroscopy in Lesson Four. You may recall that the different types of electromagnetic radiation are distinguished by varying amounts of energy, among other things. The photons produced at the sun's core are at energies much higher than those of visible light; thus, scientists must explain how these photons lose energy to produce radiation in the visible and near-visible ranges at the sun's photosphere.

By the end of this lesson it should be clear to you that many questions concerning the sun are still to be answered. Some of these questions are bound up in what has been called the "mystery of the missing neutrinos." Neutrinos are small subatomic particles, one of the "building blocks" that form the atoms themselves. They are massless, or nearly so, and therefore have the capability to move through any material at virtually the speed of light. You'll see a neutrino "telescope" that has been in use for several years in attempts to measure the sun's neutrino production. Unfortunately, far fewer neutrinos have been detected than the theoretical model predicted. The reactions of astronomers to this unexpected development are worth a few moments of reflection.

Learning Objectives

When you have completed all assignments in this lesson, you should be able to:

1. Describe and diagram the solar interior.

2. Describe the nuclear reactions occurring in the solar core.

3. Explain the radiation, conduction, and convection methods of energy transport.

4. Understand and explain how visible light is produced in the sun.

5. Describe the neutrino "problem" and its implications.

6. Name the important properties of the solar system that must be explained in any viable theory of the origin and evolution of the solar system.

7. Describe the evidence that supports the protoplanet theory of the formation of the solar system.

Assignments for This Lesson

Before Viewing the Program

Read the overview at the beginning of this lesson and look over the learning objectives that precede the assignments.

Read Chapter 13, pages 246-255 in the text.

Reread text Chapter 17. Pay careful attention to the section "The Interior of the Sun," pages 310-311. Also read carefully pages 362-367 of text Chapter 21.

View Program 15, "The Solar Interior"

As you view the program, look for:

- the explanation of how thermonuclear fusion allows the sun to produce such great amounts of energy over a long time with relatively little loss of mass.

- animation illustrating the difference between atomic fusion and fission.

- the explanation of why we see "visible" light from the sun when, in reality, invisible gamma radiation is the result of the thermonuclear fusion process in the sun's core.

- the description of the three methods of transferring energy and the two methods — radiation and convection — that are the most important on the sun.

- the description of neutrinos, the apparent "problem" between the predicted and observed amount emanating from the sun, and some possible implications of this "problem."

After Viewing the Program

Review the text assignments.

Review the terms that appear in italics and boldfaced type in the text. In particular, note the key terms and concepts listed in the study activities section in this lesson, and review those lessons and concepts listed in Lessons Four and Five.

Complete the study activities and the self-test in this study guide; then check your answers with the answer key at the end of the lesson.

The interior of the sun can be divided into three zones or areas: the core, the radiative zone, and the zone of convection. Figure 15.1 shows a cross section of the sun's interior as well as several of its exterior features.

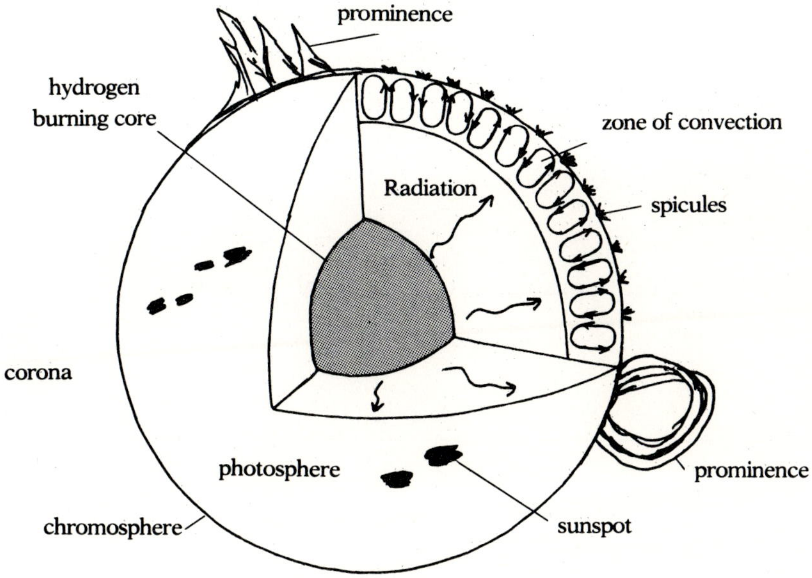

Figure 15.1. The solar interior.

Complete the following exercises:

1. Give some everyday examples of the three energy transport processes.

2. Explain how visible light is produced by the sun.

3. Without referring to the illustration in Figure 15.1, make your own diagram of the sun. Include its internal structure and atmosphere. Locate where flares, spicules, sunspots, prominences, and coronal holes occur.

Study Activities

Key Terms and Concepts

accretion
angular momentum
neutrino
neutrino "telescope"
conduction
radiation
fusion
nuclear reaction

Using What You've Learned

1. Using sources other than the text, describe the current research that attempts to explain the nature of the neutrino.

2. Make a list of the direct influences the sun has on Earth. Then try to include its indirect influences.

Recommended Readings

For supplementary reading see the bibliography in the back of this study guide.

Self-Test

(Select the one best answer.)

Objective 1: Describe and diagram the solar interior.

1. The core, radiative zone, and convective zone identify
   a. the three solar atmospheres.
   b. the three regions that constitute the solar interior.
   c. three regions within the photosphere.

Objective 2: Describe the nuclear reactions occurring in the solar core.

2. The energy in the sun is generated by
   a. carbon burning.
   b. gravity.
   c. the fusion of hydrogen into helium.
   d. unknown sources.

3. The sun's energy is generated in the
   a. photosphere only.
   b. convection zone only.
   c. entire interior.
   d. core only.

4. Light arriving on Earth today was generated by fusion reactions in the core of the sun
   a. 8 minutes ago.
   b. 8 hours ago.
   c a thousand years ago.
   d. a million years ago.

Objective 3: Explain the radiation, conduction, and convection methods of energy transport.

5. Energy carried by photons at the speed of light describes which process of energy transport?
   a. radiation
   b. convection
   c. conduction

6. The movement of matter from one location to another is necessary in the energy transport process of
   a. radiation.
   b. convection.
   c. conduction.

7. Energy moved only in the presence of matter through the vibrations of adjacent atoms in the substance describes the process of
   a. radiation.
   b. convection.
   c. conduction.

Objective 4: Understand and explain how visible light is produced in the sun.

8. Gamma and X-ray photons produced in the core eventually reach the photosphere and are emitted in the form of
   a. visible light.
   b. charged energy particles.
   c. the solar wind.
   d. the sun's magnetic field.

Objective 5: Describe the neutrino "problem" and its implications.

9. At the moment, our best chance to understand what is happening in the core of the sun lies in studying the
   a. proton.
   b. neutron.
   c. muon.
   d. neutrino.

Objective 6: Name the important properties of the solar system that must be explained in any viable theory of the origin and evolution of the solar system.

10. Which of the following must be explained by an acceptable theory of the origin and evolution of the solar system?
    a. two types of planets — small, dense, rocky ones near the sun and large gaseous/liquid ones far from the sun
    b. nearly circular orbits of the planets
    c. revolution and rotation of the planets in the same direction as the sun
    d. nearly coplanar planetary orbits
    e. all of the above

Objective 7: Describe the evidence that supports the protoplanet theory of the formation of the solar system.

11. In the development of any theory on the formation of the solar system, what properties of the solar system must be explained?
    a. differences between terrestrial and Jovian type planets
    b. formation of the sun
    c. formation of asteroids and comets
    d. all of the above

Answer Key

1. b
2. c
3. d
4. d
5. a
6. b
7. c
8. a
9. d
10. e
11. d

Surveying the Stars 16

Overview

Suppose you were driving your car slightly behind another car along a perfectly smooth and featureless highway on a rainy night. Assuming you couldn't see your car's speedometer and didn't otherwise notice any acceleration, could you tell whether your car or the car in front of you changed its speed if the distance between the two cars changed? Could you determine how fast each car was travelling if they were going at different speeds? Perhaps this problem is a little less difficult: Could you measure the diameter of a dime — if you were standing about two miles away from it? The problems faced by astronomers who seek to determine interstellar distances are not exactly like these problems, but they do have certain points in common. For example, determining the distances to even the nearest stars requires measurements as precise as would be required to measure a dime two miles away. And observations of the movement of stars must take into account that the sun, too, is moving through space.

If you're really going to understand and appreciate the distances between the stars, you should start thinking on a much grander scale. You'll recall that the sun's light takes about five hours to reach Pluto, and that it takes about four years for that light to reach the nearest star. The distances to the other stars are so great that, if you could reach Alpha Centuri, that nearest stellar neighbor, most of the constellations would appear nearly the same as they do from our solar system. The size of our entire solar system is only a fraction of a light-year; the four light-years to the next nearest star are but a fraction of the vast distance across our Galaxy. This situation is like driving from one city to the next. When you are driving between cities, you put aside the city street maps, in which one inch represents perhaps half a mile, and instead use a highway map in which an inch represents many miles. Now that you're leaving the solar system and experiencing the distances between the stars, you need to become acquainted with new units of distance, the "light-year" and "parsec."

As you study this lesson, you will become familiar with how astronomical measurements are determined. You'll discover that one surveying technique, called parallex, can be employed to determine the distance to a few stars; but the technique requires great precision and patience. In fact, some measurements, like the "proper motion" of stars, require patient observations extending over periods of years.

In this lesson you'll learn how the first distances to the stars were measured in 1838. In the century-and-a-half since then, the world has seemed to grow much smaller because of improvements in transportation and communication. And though technology has made the distant reaches of the universe easier to observe, we still perceive the universe as almost inconceivably vast.

Learning Objectives

When you have completed all assignments in this lesson, you should be able to:

1. Define and state the value of the astronomical unit.
2. List the methods for determining the astronomical unit.
3. Define the terms parallax and parsec.
4. Review and describe the historical attempts to observe stellar parallax.
5. Draw a diagram showing how stellar (trigonometric) parallax is measured.
6. Define what is meant by proper motion and by space velocity and describe how they are measured.

Assignments for This Lesson

Before Viewing the Program

Read the overview at the beginning of this lesson and look over the learning objectives that precede the assignments.

Read the introduction to Part III of the text, pages 256-258, and all of Chapter 14.

View Program 16, "Surveying the Stars"

As you view the program, look for:

- the description and visual explanation of parallax, how stellar parallax is related to the astronomical unit (AU), and how both help astronomers determine the distance to stars.
- the explanation of the "proper motion" of a star and film clips showing how the positions of stars and constellations change over spans of many years as viewed from Earth.
- visual demonstrations of how stellar motion is measured differently across the line-of-sight (by astrograph) and along the line-of-sight (by spectrograph measuring the Doppler shift) to determine its proper motion.
- the explanation and visual demonstration of how small changes or perturbations in the movement of stars can be detected and measured. These changes may indicate the presence of unseen companions, possibly other planets.

After Viewing the Program

Review the text assignment.

Review the terms that appear in italics or boldfaced type in the text. In particular, note the key terms and concepts listed in the study activities section.

Complete the study activities and the self-test in this study guide; then check your answers with the answer key at the end of the lesson.

Complete the following exercises: 1, 4, 5, 8, and 12 on page 268 in text Chapter 14.

Study Activities

Key Terms and Concepts

parallax (stellar)
astronomical unit
parsec
light-year
proper motion
radial velocity
space motion (velocity)
tangential velocity or transverse velocity

Using What You've Learned

Use the data on five hypothetical stars to answer questions 1 through 6.

	1	2	3	4	5
Distance (parsecs)	100		30		
Parallax (sec)		.0002		.04	.004
Radial Vel (km/sec)	−60	+70	+3	−3	−10
Proper Motion (sec/yr)	1.20	1.0	1.5	2.5	1.0

1. Which approaches us fastest? Which recedes from us fastest?

2. Which has a distance of 25 parsecs?

3. Which moves the fastest in space?

4. Which has a parallax of about .03 seconds?

5. Which would appear to move fastest in the sky?

6. Which is nearest and which is farthest?

Recommended Readings

For supplementary reading see the bibliography in the back of this study guide.

Self-Test

(Select the one best answer.)

Objective 1: Define and state the value of the astronomical unit.

1. The Earth-sun distance of 93 million miles is conveniently defined as one
   a. astronomical unit.
   b. parsec.
   c. light-year.
   d. solar system constant.

Objective 2: List the methods for determining the astronomical unit.

2. Measuring the distance to another planet by radar bounce and using Kepler's third law is a technique used to determine the value of the astronomical unit.
   a. true
   b. false

Objective 3: Define the terms parallax and parsec.

3. About how far away (in parsecs) is a star that is 30 light-years away?
   a. 10
   b. 15
   c. 200
   d. 3,000

4. Alpha Centauri is at a distance of 1.35 parsecs and has the largest stellar parallax. Its parallax is about
   a. 1 arc second.
   b. .75 arc second.
   c. .03 arc second.
   d. 1.4 arc second.

Objective 4: Review and describe the historical attempts to observe stellar parallax.

5. The first stellar parallax was measured in the
   a. sixteenth century.
   b. seventeenth century.
   c. eighteenth century.
   d. nineteenth century.
   e. twentieth century.

Objective 5: Draw a diagram showing how stellar (trigonometric) parallax is measured.

6. The maximum distance that can be measured by the method of stellar parallax is 100 parsecs. What is the parallax of a star at that distance?
   a. 0.1 arc second
   b. 1.0 arc second
   c. .01 arc second
   d. .001 arc second

7. If you observed the parallax of a star from a moon of Saturn, 10 AU from the sun, it would be
   a. ten times greater than as observed from Earth.
   b. 100 times greater than as observed from Earth.
   c. one-tenth as great as observed from Earth.
   d. the same as observed from Earth.

Objective 6: Define what is meant by proper motion and by space velocity and describe how they are measured.

8. If you know the radial and tangential velocity of a star and its distance, you can determine the star's
   a. rotation.
   b. space velocity.
   c position.
   d. age.

9. Proper motion of the stars will cause the constellations to
   a. fade away in a few years.
   b. change their appearance over very long periods of time.
   c. appear closer and therefore brighter.
   d. remain the same.

10. ____________ is the nearest star and ____________ has the largest proper motion.
   a. Barnard's star, Alpha Centauri
   b. Sirius, Beta Lyrae
   c. Alpha Orionis, Zeta Reticuli
   d. Alpha Centauri, Barnard's star

11. What is measured by photographing a star on two different nights separated by many years, measuring the change in position from one photograph to another, then correcting for parallax?
   a. the Doppler shift in the stellar spectrum
   b. radial velocity
   c. proper motion
   d. rotation of Earth

Answer Key

1. a
2. a
3. a
4. b
5. d
6. c
7. a
8. b
9. b
10. d
11. c

The Message of Starlight 17

Overview

In the previous lesson, you learned that parallax has some limitations when used to measure the distances to stars. Accurate measurement is possible only in our nearby stellar neighborhood, up to about 60 light-years. Although this neighborhood takes in a lot of space, compared to the size of our own solar system, it's not too significant if astronomers are going to measure the distances to other stars within our own Galaxy (which, remember, is some 100,000 light-years in diameter).

If all the stars in our Galaxy were about the same brightness as our own sun, finding distances wouldn't be too great a problem. Brightness decreases according to the square of the distance, so brighter-appearing objects would be closer and dimmer objects farther away. The distance itself could be determined with considerable accuracy using sensitive photoelectric devices to measure the amount of light received. Our sun may be considered average, but the suns in our Galaxy exist in a wide range of sizes, colors, and brightness. In fact, some suns would hardly be very visible to our eyes, because much of their energy is emitted in the invisible portions of the electromagnetic energy spectrum.

Fortunately, the laws of nature apply throughout the universe, and there are interrelationships among a star's size, temperature, composition, brightness, and spectrum. Stars send messages to our astronomers, messages that identify the properties of those stars. In the past one hundred years or so, scientists have been decoding these messages with more and more success. This lesson is really an introduction to part of the code. But it should be thorough enough that you can understand how spectrograms of a point of light (the way all stars appear in even the largest telescopes) can reveal information about a star's magnetic field, rate of rotation, or chemical composition.

In this lesson, you'll learn how photometry, or the study of light, has advanced beyond visual and photographic techniques. You'll learn about a system for rating brightness on a numerical or magnitude scale, and discover that for every star there are several different types of magnitudes important to the astronomer. Stars can be described as belonging to a particular spectral class. The spectral class of a star tells the astronomer much about its temperature and actual brightness, and therefore provides a basis for determining the star's distance.

The lesson also introduces you to several types of variable stars. Variable stars do not emit energy at a uniform rate. This variation of energy output enables astronomers to learn a great deal about the life cycles of stars and, in the case of cepheid and RR Lyrae variables, their distances.

Learning Objectives

When you have completed all assignments in this lesson, you should be able to:

1. List and describe the important types of photometry.
2. Define the terms apparent magnitude, absolute magnitude, luminosity, color index, and bolometric magnitude.
3. Calculate how much brighter one star is than another when the magnitudes of both are given.
4. Calculate the color index of a star when the blue and visual magnitudes are given.
5. Describe the different luminosity classes of stars.
6. List the spectral classes of stars in order of temperature and describe their characteristics.

7. List the stellar properties that can be determined from stellar spectrograms, and describe how these properties are deduced.

8. Contrast the relative sizes, temperatures, and luminosities of stars in different regions on an H-R diagram.

Assignments for This Lesson

Before Viewing the Program

Read the overview at the beginning of this lesson and look over the learning objectives that precede the assignments.

Read all of text Chapter 15, and Chapter 16, pages 283-285 and 292-296.

View Program 17, "The Message of Starlight"

As you view the program, look for:

- the explanation and measurement of the magnitude scale for stellar brightness and how this information can be used to determine stellar distances.

- the description of variable stars and their use as distance indicators.

- the explanation of how the spectrum of a star can reveal its chemical composition, temperature, motion, and other properties.

- the classification of stars according to their spectrum and how this information is used, along with stellar brightness, to construct the Hertzsprung-Russell diagram.

- the description of the various groupings of stars on the H-R diagram (main sequence, red giant, white dwarf).

After Viewing the Program

Review the text assignments.

Review the terms that appear in italics or boldfaced type in the text. In particular, note the key terms and concepts listed in the study activities section.

Complete the study activities and the self-test in this study guide; then check your answers with the answer key at the end of the lesson.

After reading the section on spectral classification, you probably realized that two stars could have the same spectral class (same temperature), M, for example. However, because the two stars have the same surface temperature — M class is about 3,500°K — they may not necessarily be the same kind of star. One star could be a red giant, hundreds of millions of miles in diameter, and one could be a main sequence red dwarf, a little smaller than our sun.

This confusion is remedied by adding an additional classification code. This code represents "luminosity class." It identifies whether the star is a supergiant, giant, or main sequence star. The code is as follows:

I. bright supergiants
II. bright giants
III. giants
IV. subgiants
V. main sequence, dwarf stars

This Roman numeral notation is usually added to the spectral classification of the star. See text Appendicies 14 and 15.

Complete the following exercises: 3, 7, 9, and 11 in text Chapter 15, pages 281-282.

Study Activities

Key Terms and Concepts

apparent magnitude
absolute magnitude
photometry
bolometric magnitude
variable star
cepheid variable
RR Lyrae variable
spectral type (class)
Hertzsprung-Russell diagram
main sequence
red giant
white dwarf
Zeeman effect
Doppler effect
luminosity class
color index
luminosity

Using What You've Learned

1. Using a piece of graph paper, plot a Hertzsprung-Russell diagram for the stars listed in text Appendices 14 and 15, the nearest and brightest stars, respectively. You will be plotting the spectral type of the A component with the absolute visual magnitude. The star Sirius (spectral type A1 and absolute visual magnitude of the A component +1.4) and our sun (G2; +4.6) are already plotted. Using Figure 16.9 on text page 293 for guidance, try to answer the following questions:

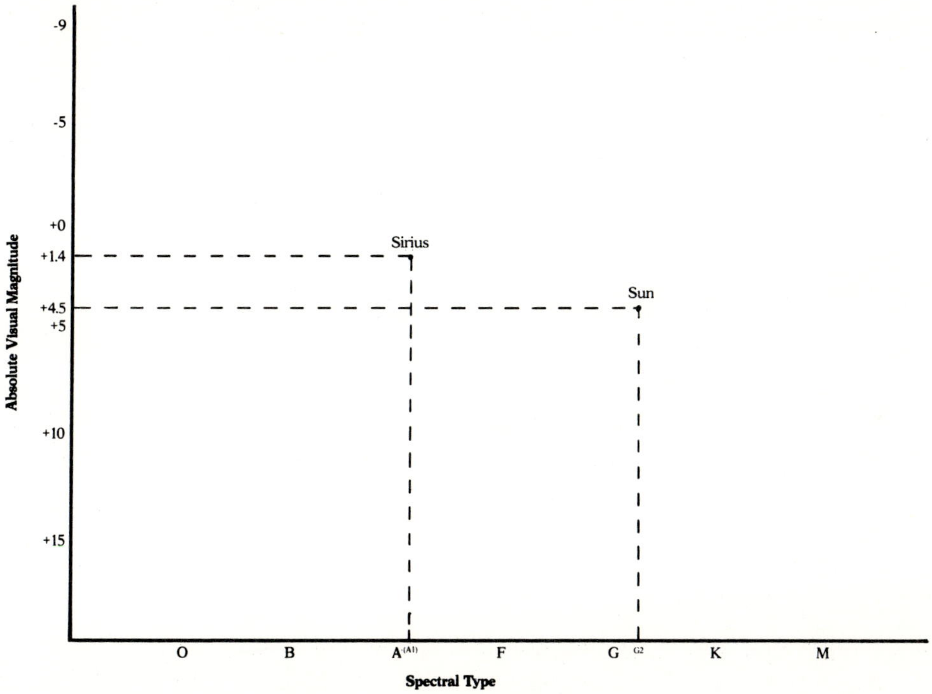

Figure 17.1. Graph of H-R diagram.

a. Identify the main sequence, red giant, and white dwarf branches of the diagram.

b. Briefly describe the differences in position of the H-R diagram of the nearest and brightest stars.

c. Once you have plotted the 20 brightest stars, use the right ascension and declination coordinates in text Appendix 15 and locate these stars on the star chart found in Lesson Eight on Mercury and Venus (you'll find 18 out of 20). Right ascension and declination have the same use in space as do longitude and latitude, respectively, on the surface of Earth. Label the star with its H-R diagram branch (main sequence, red giant, etc.).

d. Go outside and try to find these special stars in the night sky. Check the seasons when these stars are visible.

2. Consider the following H-R diagram. Which star is:

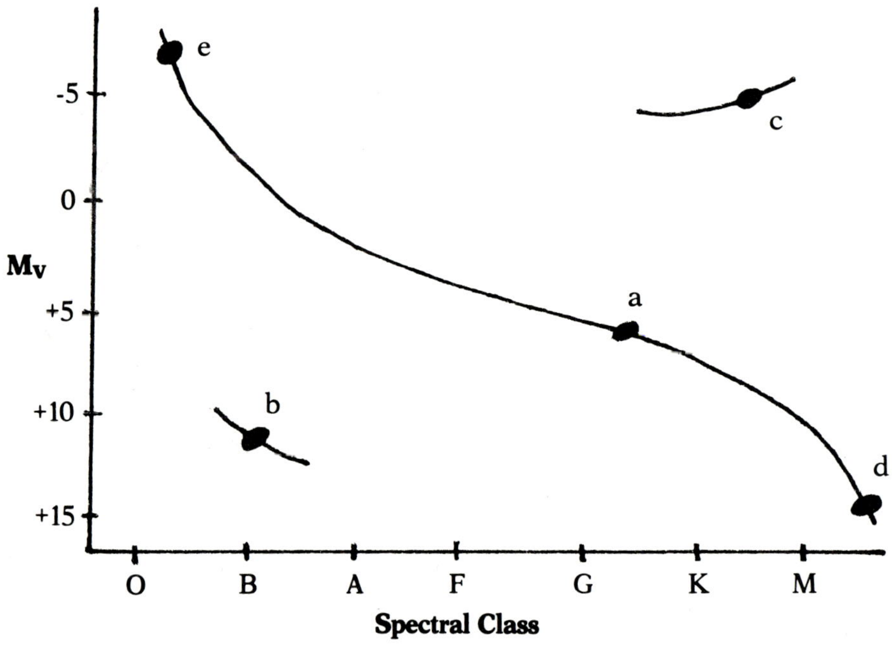

Figure 17.2. H-R diagram.

a. the largest in size?
b. the smallest in absolute magnitude?
c. smallest in size?
d. most common?
e. like the sun?
f. most advanced in its evolution?
g. the hottest?
h. certainly still young?
i. a white dwarf?
j. a red dwarf (a faint red main sequence star)?

Recommended Readings

For supplementary reading see the bibliography in the back of this study guide.

Self-Test

(Select the one best answer.)

Objective 1: List and describe the important types of photometry.

1. Measuring the size of an image and its darkness on photographic negatives is the technique used in ____________ photometry.
   a. photoelectric
   b. magnitude
   c. photographic
   d. spectroscopic

Objective 2: Define the terms apparent magnitude, absolute magnitude, luminosity, color index, and bolometric magnitude.

2. According to the original magnitude system of Hipparchus, a(n) ____________ magnitude star is the faintest that can be seen by the unaided human eye.
   a. first
   b. third
   c. sixth
   d. eighth

Objective 3: Calculate how much brighter one star is than another when the magnitudes of both are given.

3. Barnard's star is about a tenth magnitude star and Tau Ceti is about a fourth magnitude star. Which star is brighter and by how many times?
   a. Barnard's star is about six times brighter.
   b. Barnard's star is about fifteen times brighter.
   c. Tau Ceti is 100 times brighter.
   d. Tau Ceti is 250 times brighter.

4. If you move a star six times farther away, how many times fainter will it appear?
   a. 6 times
   b. 12 times
   c. 36 times
   d. 100 times

Objective 4: Calculate the color index of a star when the blue and visual magnitudes are given.

5. If star A has color magnitudes B = 4 and V = 3, and star B has color magnitudes B = 2 and V = O, which star is *bluer* in color?
   a. star A
   b. star B
   c. The two stars are the same color.
   d. The answer cannot be determined from the information given.

6. Two stars have the same absolute visual magnitude of $^{+}2$. If star A is yellow in color, what must be the color of star B in order for it to have a smaller (brighter) absolute bolometric magnitude?
   a. blue
   b. yellow
   c. orange
   d. red

Objective 5: Describe the different luminosity classes of stars.

7. Star A is a G3 V. Which of the following stars is larger and more luminous than star A?
   a. K3 V
   b. G3 III
   c. M2 I
   d. M3 V

Objective 6: List the spectral classes of stars in order of temperature and describe their characteristics.

8. A G3 II star has a higher surface temperature than a
   a. K4 I.
   b. B3 II
   c. AO III
   d. GO V.

9. A star of surface temperature 10,000° K would be classified
   a. B3.
   b. AO.
   c. G2.
   d. M5

10. The strongest hydrogen lines in the visible part of the spectrum are exhibited by a star of spectral type A.
    a. true
    b. false

Objective 7: List the stellar properties that can be determined from stellar spectograms, and describe how these properties are deduced.

11. Which property of the star can possibly be determined from its spectrum?
    a. temperature
    b. composition
    c. radial velocity
    d. all of the above

Objective 8: Contrast the relative sizes, temperatures, and luminosities of stars in different regions on an H-R diagram.

12. A main sequence star would have a temperature of ____________ and an absolute magnitude of ____________.
    a. 20,000° K, +15
    b. 6,000° K, −10
    c. 6,000° K, +5
    d. 2,800° K, −8

13. Of the stellar properties listed, which ones could be used to construct an H-R diagram?
    a. mass
    b. surface temperature
    c. radial velocity
    d. luminosity

Answer Key

1. c	8. a
2. c	9. b
3. d	10. a
4. c	11. d
5. a	12. c
6. a	13. b, d
7. b, c	

Binary Stars 18

Overview

During November, December, and January there's a bright second magnitude star nearly overhead. This star, which is easily observed in the northern hemisphere, exhibits some strange behavior with great regularity. Every 2.867 days it dims from its usual brightness to a magnitude of about 3.4, then brightens again, the entire change taking about 10 hours. The star, Algol, is a binary with a small companion, and the plane of their orbits lies in Earth's line of sight, so that we can observe the two stars eclipse each other. This binary, as you'll find out in this lesson, isn't a rarity in this star-filled Galaxy. Actually, there are more binary stars than single ones, making our sun a member of the less numerous group.

It may seem strange that an entire lesson is devoted to binary stars, even if they are in the majority. The reason for such attention is that when two bodies orbit each other, astronomers can apply Newton's law of gravity and determine the combined masses of the bodies. In fact, it's sometimes possible to determine the individual mass of each body. Thus scientists have been able to take a sampling of the masses of a number of stars in the Galaxy and draw some important conclusions about stellar evolution and energy output from their data.

In this lesson, the complex observations and calculations that determine the orbits and the masses of "visual binaries" (those that can be seen as individual stars by a telescope) will be explained to you; some of these procedures will be shown in the television program, as well. Many binary stars can't be identified individually by telescope, but astronomers can spot some of these through clues in their spectra. The third type of binary star is the eclipsing binary. Eclipsing binaries have yielded even more information, including stellar radii.

In the television program for this lesson, you'll view one of the first pictures of a star's disk taken through computer-processing of split-second images called "speckle interferometry." This technique was mentioned briefly in Lesson Five and now is described by one of those who developed the process. In the future we may see more images of stars, both from this process and from the greater resolving power of telescopes placed in orbit above Earth's atmosphere.

Studies of the mass of stars have enabled astronomers to identify significant relationships between mass and energy output (luminosity). There also appears to be a limit to the amount of matter a star can contain.

Learning Objectives

When you have completed all assignments in this lesson, you should be able to:

1. List the major types of binary stellar-systems and describe their characteristics.
2. Describe the stellar mass-luminosity relation.
3. Describe the methods for determining the sizes and masses of binary stars.
4. Describe the apparent mass limit for stars.
5. Describe the "speckle photography" method of obtaining stellar images.
6. Describe the luminosity function.

Assignments for This Lesson

Before Viewing the Program

Read the overview at the beginning of this lesson and look over the learning objectives that precede the assignments.

Reread text Chapter 16, pages 283-285. Read carefully pages 285-292 and review the rest of Chapter 16, pages 292-297.

View Program 18, "Binary Stars"

As you view the program, look for:

- the series of diagrams identifying the barycenter of a binary star system and illustrating how it is located and how it and the orbital sizes and periods of the stars are used to determine their respective masses.
- demonstrations of how the apparent distance (angular separation) between two binary components is measured and adjusted mathematically to determine true orbital dimensions.
- the explanation of how the study of eclipsing binary star systems can enable us to obtain the diameters of individual stars.
- the explanation of how the atmosphere distorts visible starlight when viewed at Earth's surface and how the techniques of "speckle interferometry" are used to achieve better resolution of a star's light.
- the explanation of why the largest (red giant) and smallest (white dwarf) stars differ far more with respect to their sizes than to their masses; and the description of the "basic mass-luminosity" relationship of most stars.

After Viewing the Program

Review the text assignments.

Review the terms that appear in italics or boldfaced type in the text. In particular, note the key terms and concepts listed in the study activities section.

Complete the study activities and the self-test in this study guide; then check your answers with the answer key at the end of the lesson.

Complete the following exercises: 6 and 9 in text Chapter 16, pages 296-297.

Study Activities

Key Terms and Concepts

luminosity function
binary star
visual binary
spectroscopic binary
eclipsing binary
mass-luminosity relation
speckle interferometry

Using What You've Learned

Use the data given here on five hypothetical stars to answer questions 1 through 11.

Star	m	M	Parallax	Distance (parsecs)	Spectral Type
a	5			10	main sequence G
b	2.1		.01		M giant
c	3.7	3.7			main sequence F
d	13				main sequence O
e	−1.4	1.4	.38		main sequence A

1. Which is 100 pc away?
2. Which is like the sun?
3. Which appears brightest in the sky?
4. Which is intrinsically the brightest?
5. Which is the nearest?
6. Which is most distant?
7. Which is largest in radius?
8. Which is hottest?
9. Which appears faintest in the sky?
10. Which is coolest?
11. Which has the numerically largest absolute magnitude?

Recommended Readings

For supplementary reading see the bibliography in the back of this study guide.

Self-Test

(Select the one best answer.)

Objective 1: List the major types of binary stellar systems and describe their characteristics.

1. Visual binary stars have the longest average period of revolution.
   a. true
   b. false

2. Spectroscopic binary systems are found, on the average, closer to Earth than visual binary systems.
   a. true
   b. false

3. About half of the nearest stars to our sun are multiple star systems.
   a. true
   b. false

4. The period of revolution of a binary system can be
   a. a few hours.
   b. a few months.
   c. thousands of years.
   d. all of the above.

Objective 2: Describe the stellar mass-luminosity relation.

5. The mass-luminosity relationship is true for
   a. only main sequence stars.
   b. 80% of giant stars.
   c. 90% of all stars.
   d. only white dwarfs.

6. If the mass of a star is tripled, the luminosity would
   a. increase about 10 times.
   b. decrease about 3 times.
   c. increase about 50 times.
   d. decrease about 30 times.

Objective 3: Describe the methods for determining the sizes and masses of binary stars.

7. Given the right conditions, the eclipsing binary enables the astronomer to determine stellar
   a. temperatures.
   b. diameters.
   c. distances.
   d. composition.

Objective 4: Describe the apparent mass limit for stars.

8. The most massive stars are ____________ times the mass of the sun and the least massive stars are ____________ times the mass of the sun.
   a. 50, .01
   b. 1,000, 1
   c. 500, .1
   d. 10, .001

9. In order to find out the mass of each member in a binary system, which of the following must be determined?
   a. position of the barycenter
   b. period of revolution
   c. average distance between the stars
   d. all of the above

Objective 5: Describe the "speckle photography" method of obtaining stellar images.

10. Which of the following techniques have been used to measure stellar diameters?
    a. stellar interferometer
    b. speckle interferometry
    c. high-speed photoelectric observations
    d. all of the above

Objective 6: Describe the luminosity function.

11. According to the luminosity function, the majority of stars in the solar neighborhood are
    a. more luminous than the sun.
    b. less luminous than the sun.
    c. about the same luminosity as the sun.

Answer Key

1. a
2. b
3. a
4. d
5. c
6. c
7. b
8. a
9. d
10. d
11. b

The Milky Way — Discovered 19

Overview

Every lesson of **Project Universe** recounts one or more stories of significant discoveries in astronomy. In "The Milky Way — Discovered," you will retrace some of the steps that you have taken in order to grasp the grand design of the Galaxy. You, in turn, will discover that the overall structure of the Milky Way galaxy and the universe beyond are nearly hidden from sight, obscured not only by distance but by clouds of dust and gases thin enough to be a vacuum, yet substantial enough to hide much of the light in the universe from us.

In this lesson, you will learn the methods used by several astronomers to discover the overall shape of this huge collection of stars called our Galaxy. To appreciate the scope of this problem and better understand why the description of the structure of the Galaxy was controversial for a time, you'll learn some more about the properties of the cepheids and other variable stars. You also will be introduced to dust and gas nebulas and the tenuous invisible matter that fills the spaces between the stars, and therefore creates special problems, such as interstellar reddening, for astronomers. You'll learn about the globular clusters, groups of stars which are distributed around the Galaxy in such a way that they enabled Harlow Shapley to locate the position of the sun within the Galaxy. In the television program you will see a reenactment of the painstaking observations which enabled Edwin Hubble to identify the true nature of a hazy spot in Andromeda. That spot turned out to be another galaxy, some two million light-years distant, similar to but totally separate from the Milky Way.

The Andromeda galaxy, by the way, is perceptible by the naked eye on a clear dark night during late fall and early winter. It's described on many star maps as "M31," the thirty-first object listed by Charles Messier in his 1781 catalog of hazy objects. Messier's catalog carefully described over 100 patches of light in the sky, which later were recognized as galaxies, star clusters, and nebulas. Unfortunately, because Messier was looking only for comets, he identified all these objects to keep them from being confused with new comets and didn't bother to find out what the objects actually were.

Learning Objectives

When you have completed all assignments in this lesson, you should be able to:

1. Describe the role of the observations of Herschel and Shapley in determining the size of the Milky Way and the sun's position within it.

2. List and describe the different types of nebulas.

3. Describe the conditions prevailing in HI and HII regions.

4. Name the contents of the interstellar regions.

5. List and describe the properties of globular clusters.

6. Define and describe the light curves for cepheid variables.

7. Draw the period-luminosity relation and explain how it is used in distance measurements.

8. Explain the importance of the 21-cm line of neutral hydrogen.

Assignments for This Lesson

Before Viewing the Program

Read the overview at the beginning of this lesson and look over the learning objectives that precede the assignments.

Read all of text Chapter 18 and pages 411-416 in Chapter 24.

Reread the section on types of variable stars in text Chapter 15, pages 273-275.

View Program 19, "The Milky Way – Discovered"

As you view the program, look for:

- the development of our understanding of the shape and content of the Milky Way galaxy (including the work of Herschel, Kapteyn, and Shapley).
- the description of the distribution and content of interstellar gas and dust in the Milky Way and their effects on starlight.
- the explanation of the period-luminosity relation for variable stars and how this information is used to determine distances.
- the description of and differences between RR Lyrae, cepheid, and Mira-type variable stars.
- the nature and distribution of globular clusters.
- the description of HI and HII regions.

After Viewing the Program

Review the text assignments.

Review the terms that appear in italics or boldfaced type in the text. In particular, note the key terms and concepts listed in the study activities section.

Complete the study activities and the self-test in this study guide; then check your answers with the answer key at the end of the lesson.

Toward the end of the nineteenth century, Jacobus Kapteyn conducted a comprehensive study of the distribution of stars. He concluded that the Milky Way galaxy was 50,000 light-years wide and 6,000 light-years thick, with the sun located at the center of the Galaxy. Kapteyn published his study in 1922. Ironically, five years earlier, Harlow Shapley arrived at a model of the Galaxy that is very much like the model we have today, although the dimensions have changed.

Complete the following exercises: 1, 2, 4, 5, and 8 in text Chapter 18, page 331.

Study Activities

Key Terms and Concepts

interstellar dust
dark nebulas
globules
reddening (interstellar)
color excess
reflection nebulas
HI regions
HII regions
21-cm line of hydrogen
galaxy
Milky Way galaxy
Andromeda galaxy
Clouds of Magellan
globular clusters
cepheid variable
period-luminosity relation

Using What You've Learned

1. Draw a side view of our Galaxy indicating the nucleus, the sun's position, the diameter, and a few globular clusters.

2. In what way might it be likely that HI and HII regions could be related to star formation?

3. Why might a star behind a dust cloud appear reddish (even if it weren't a red star)?

Recommended Readings

For supplementary reading see the bibliography in the back of this study guide.

Self-Test

(Select the one best answer.)

Objective 1: Describe the role of the observations of Herschel and Shapley in determining the size of the Milky Way and the sun's position within it.

1. By examining the distribution of globular clusters in and near our Galaxy, Harlow Shapley discovered
   a. that the actual shape of the Milky Way was spherical.
   b. the sun's off-centered position in our Galaxy.
   c. that the Milky Way contained over 100 billion stars.
   d. the total count of globular clusters belonging to our Galaxy.

2. Our sun is located approximately how many parsecs from the center of our Galaxy?
   a. 1,000
   b. 9,000
   c. 18,000
   d. 30,000

Objective 2: List and describe the different types of nebulas.

3. A reflection nebula would most likely show a (n)
   a. continuous spectrum.
   b. emission spectrum.
   c. bright line spectrum.
   d. absorption spectrum.

Objective 3: Describe the conditions prevailing in HI and HII regions.

4. An emission nebula of hot ionized hydrogen, reddish in color, best describes a (n)
   a. planetary nebula.
   b. HII region.
   c. reflection nebula.
   d. HI region.

Objective 4: Name the contents of the interstellar regions.

5. Dust grains in the interstellar medium are primarily responsible for the
   a. reddening and dimming of starlight.
   b. Doppler effect.
   c. brightening of starlight.
   d. spectral wavelength shifts toward the blue end of the spectrum.

Objective 5: List and describe the properties of globular clusters.

6. Globular clusters may contain one hundred thousand stars.
   a. true
   b. false

Objective 6: Define and describe the light curves for cepheid variables.

7. The light curve of a variable star continuously changes in brightness from brightest to dimmest with a period of many days. This star is most likely a
   a. Mira-type variable.
   b. RR Lyrae variable.
   c. cepheid variable.

8. By relating the period of a cepheid to its luminosity and using it as a standard candle, the cepheid's ______________ can be determined.
   a. age
   b. mass
   c. size
   d. distance

Objective 7: Draw the period-luminosity relation and explain how it is used in distance measurements.

9. The longer the period of a cepheid, the greater its luminosity. This statement is based on the
   a. period-luminosity relation.
   b. mass-luminosity relation.
   c. Hertzsprung-Russell diagram.
   d. temperature-luminosity relation.

Objective 8: Explain the importance of the 21-cm line of neutral hydrogen.

10. The 21-cm line of hydrogen helped astronomers to locate the
    a. globular clusters.
    b. spiral arms of the Galaxy.
    c. center of the Galaxy.
    d. position of the sun within the Galaxy.

Answer Key

1. b
2. b
3. d
4. b
5. a
6. a
7. c
8. d
9. a
10. b

The Milky Way — Structure 20

Overview

On the darkest of summer nights, you could turn your eyes in the direction of the Milky Way's nucleus — the most densely crowded and brightest part of our Galaxy — and see absolutely no sign of it. The galactic nucleus, as you'll learn in this lesson, lies in the direction of the constellation of Sagittarius, which we see about 30 degrees south of the celestial equator. However, the galactic nucleus is completely hidden from our eyes (and will always be hidden) by the dust and gas which lie in our line of sight. Photographs of neighboring systems like the Andromeda galaxy give us, perhaps, the best suggestion of the vast panorama of our own Galaxy.

Fortunately, as you've already learned in previous lessons, the spectrum of electromagnetic radiation is much more extensive than what we can see with our eyes. And for reasons that will be mentioned later in this lesson, the Galaxy is much more transparent to the longer-wavelength, lower-energy radiation detected as radio and infrared energy. We aren't totally blindfolded in our own Galaxy, but the messages received from visible light, "radio light," or other radiation become more and more difficult to decipher as astronomers reach farther into space.

You may find it helpful to think of the Milky Way galaxy from two separate perspectives as you learn what is known today about its structure. First, the Galaxy is made up of many different classes and subclasses of objects and other matter. In the last lesson, for example, you learned about HI and HII regions. In this lesson, you'll learn about Population I and Population II stars and different kinds of star clusters. So, one way to look at the Galaxy is as a collection of a large number of different kinds of objects.

A second perspective on the Galaxy is to view it as a whole. Its stars are bound together by gravity. Its regions — the nucleus, the disk, the spiral arms, and the halo — have distinctive features such as the distribution and content of materials and types of stars. In a way, such a "double perspective" of this Galaxy should be no more difficult than the use of a modern highway atlas in planning an automobile trip across the continent. As mentioned in the previous lesson, sometimes you'd use the map of the entire continent to get an overall picture, while other times you'd use maps of individual states or even cities to locate specific streets or areas.

During the course of this lesson, you'll be investigating many of the features of the Milky Way. You'll observe how radio maps of the Galaxy are obtained and learn how they provide clues to the actual design of the Galaxy. You'll find out how the sun's orbit around the nucleus has been studied and how this study provided a basis for computing the total mass of the Galaxy. And you'll learn how the Galaxy itself came into being — and how our sun is a kind of "junior member," born sometime after the first generation of stars.

Learning Objectives

When you have completed all assignments in this lesson, you should be able to:

1. Draw a diagram of the Milky Way and name the major regions, including the spiral arms near the sun.

2. Describe and name the sun's galactocentric velocity and period.

3. Describe how the mass of the Milky Way is determined.

4. Distinguish between Population I and II type stars.

5. List and describe the properties of galactic clusters.

6. Describe the radio mapping of the Milky Way.

7. Describe what causes the spiral arms.

8. Describe the formation of the Milky Way.

9. Describe the nature, content, and mass of the galactic halo.

10. Describe the behavior of the galactic nucleus.

Assignments for This Lesson

Before Viewing the Program

Read the overview at the beginning of this lesson and look over the learning objectives that precede the assignments.

Read all of Chapter 24 in the text. Also read pages 379-384 of Chapter 22.

View Program 20, "The Milky Way – Structure"

As you view the program, look for:

- the description of the differential rotation of the Milky Way and the location of our sun in it.
- the description of the discovery of the 21-centimeter line of neutral hydrogen and its use in determining the spiral structure of the Milky Way.
- the description and distribution of galactic clusters.
- the description and distribution of Population I and Population II type stars.
- the explanation of the formation of the spiral arms and the formation of the entire galaxy itself.
- the description and content of the galactic halo.

After Viewing the Program

Review the text assignment.

Review the terms that appear in italics or boldfaced type in the text. In particular, note the key terms and concepts listed in the study activities section.

Complete the study activities and the self-test in this study guide; then check your answers with the answer key at the end of the lesson.

One theory that explains the continuance of the spiral arms is the density wave theory. This theory states that density or compression waves move through the interstellar medium, possibly as a result of gravitational differences in the disk. These waves do not consist of moving matter. They move outward from a center at a speed slower than the interstellar clouds. As these clouds enter a compressional wave, matter is compressed to a high enough density to initiate star formation. The continued motion of the density wave and stellar matter create the spiral pattern we observe in our Galaxy.

Complete the following exercises: 4, 5, 6, 8, and 10 in text Chapter 24, pages 424-425.

Study Activities

Key Terms and Concepts

nuclear bulge
spiral arms
halo (of a galaxy)
Population I and II
synchrotron radiation
thermal radiation
nonthermal radiation
open clusters
associations (of stars)
Sagittarius A
density wave

Using What You've Learned

In answering questions 1 through 5, you may choose among the following:

a. nuclear bulge
b. disk
c. spiral arms
d. corona (or halo)
e. globular clusters

1. Which gave Shapley the clue to the size of the Galaxy?

2. Which contains the lowest density of stars?

3. Which contains the highest concentration of the Galaxy's mass?

4. Which is where most of the interstellar matter is concentrated?

5. Which is where young luminous stars can be found?

6. List the evidence that supports the fact that the Milky Way has a magnetic field.

Recommended Readings

For supplementary reading see the bibliography in the back of this study guide.

Self-Test

(Select the one best answer.)

Objective 1: Draw a diagram of the Milky Way and name the major regions, including the spiral arms near the sun.

1. The spiral arms contain
   a. the highest density of stars.
   b. most of the gas and dust of our Galaxy.
   c. most of the globular clusters.
   d. nothing — they're an illusion.

Objective 2: Describe and name the sun's galactocentric velocity and period.

2. The sun takes approximately how many years to orbit the center of the Galaxy?
   a. 20
   b. 2,000
   c. 2,000,000
   d. 200,000,000

Objective 3: Describe how the mass of the Milky Way is determined.

3. At present there are ________________ stars in the Milky Way galaxy.
   a. less than a billion
   b. more than 10 trillion
   c. over 100 billion
   d. an infinite number of

4. The total number of stars in the Milky Way was determined by counting the number of stars in selected regions and then estimating the number of those regions in the Galaxy.
   a. true
   b. false

Objective 4: Distinguish between Population I and II type stars.

5. Extreme Population I stars have the
   a. weakest concentration of chemical elements heavier than hydrogen and helium.
   b. weakest concentration of chemical elements heavier than iron.
   c. highest concentration of chemical elements heavier than hydrogen and helium.
   d. highest concentration of chemical elements lighter than hydrogen and helium.

Objective 5: List and describe the properties of galactic clusters.

6. The stars of galactic clusters are
   a. Population II type stars.
   b. the oldest stars in the Galaxy.
   c. the youngest, most recently formed stars.
   d. stars that were once located in the central bulge of the Galaxy.

Objective 6: Describe the radio mapping of the Milky Way.

7. The 21-cm line helped astronomers to map the spiral arms of the Galaxy. The 21-cm line originates from
   a. galactic clusters.
   b. globular clusters.
   c. HII regions.
   d. HI regions.

Objective 7: Describe what causes the spiral arms.

8. The spiral structure of the Milky Way is a relatively permanent feature identified by the presence of O and B type stars and concentration of neutral hydrogen.
   a. true
   b. false

Objective 8: Describe the formation of the Milky Way.

9. The stars in globular clusters
   a. are believed to have formed earliest in the history of our Galaxy.
   b. were once located in the spiral arms and have recently moved to their present location.
   c. are the hottest stars in our Galaxy.
   d. are Population I type stars.

Objective 9: Describe the nature, content, and mass of the galactic halo.

10. When the mass of the galactic halo is included, the total mass of the Milky Way is estimated to be
    a. a billion suns.
    b. a million suns.
    c. more than a trillion suns.
    d. an infinite number of suns.

Objective 10: Describe the behavior of the galactic nucleus.

11. The nucleus of the Milky Way is emitting radiation ranging from gamma rays to radio waves.
    a. true
    b. false

Answer Key

1. b	7. d
2. d	8. a
3. c	9. a
4. b	10. c
5. c	11. a
6. c	

Extraterrestrial Communication 21

Overview

Is anybody out there? In the last decade or so, the popularity of certain films and television programs gives some indication that there is hope for a positive answer to this question. Indeed, people seemed to be disappointed at the results of the Viking experiments in the late 1970s, which gave no positive evidence of life on Mars. Despite years of scientific (and pseudoscientific) studies of UFOs, there is no scientific evidence for intelligent life, or even of rudimentary cellular life as we know it, anywhere else in the universe. There is some evidence, discussed in the text and the television program, that life based on chemistry similar to that on Earth could occur under the right conditions. There have been serious efforts to calculate the probability of the existence of intelligent life, but, as you will learn, there are some unknown factors that make such calculations imprecise or of very little value.

While you consider the information presented in this lesson, you might wish to ask yourself what priority you would assign to a search for extraterrestrial intelligence (ETI). There is always a limit to the money that can be invested in astronomical research over any given period. In the preceding lessons, you probably noticed a number of questions about astronomy awaiting further study. Would the time and finances available be better devoted to seeking answers for some of these questions, rather than to searching for signs of a civilization elsewhere in space?

If communication with another civilization were successfully established, even one-way communication, what might it mean to us? Might it supply human cultures with new solutions to the seemingly insurmountable problems faced today? Even if the probability of achieving some kind of contact is very poor, are the possible gains worth the gamble? You may have already guessed that this lesson does not answer any of these questions, any more than it can reveal whether or not life exists other than on our planet. But you will be given some data and some hypotheses that keep interest in ETI alive in the United States and other countries today.

In this lesson, you will learn how life may have begun on Earth and about the conditions which should be optimum for life to begin on other planets. Some serious searches for radio signals from distant civilizations have been carried out. Searches made by the United States include Project Ozma, Project SETI, and Project Cyclops, all of which are described in the television program. Other searches have been proposed by the U.S. and other countries. There have also been attempts to send messages into space, such as the message sent by the Arecibo radio telescope. Perhaps you'll find yourself sharing the enthusiasm of those who impatiently champion an expanding program of radio searches. Maybe you'll be more concerned with the problems Earth's civilization faces. No matter what your feelings are, you'll probably finish your study of this lesson with a new awareness that Earth's civilization might not be the only one in the universe.

Learning Objectives

When you have completed all assignments in this lesson, you should be able to:

1. List the criteria involved in calculating the number of communicative civilizations in the Galaxy.

2. Describe the attempts made by astronomers to detect planets in orbit around other stars.

3. Describe the modern attempts to contact extraterrestrial civilizations.

4. Describe the conditions necessary for a solar system to contain planets with the possibility of life evolving on them.

Assignments for This Lesson

Before Viewing the Program

Read the overview at the beginning of this lesson and look over the learning objectives that precede the assignments.

Read all of Chapter 28 in the text.

View Program 21, "Extraterrestrial Communication"

As you view the program, look for:

- an explanation of what is meant by a "life zone" around a star.
- a description of the Stanley Miller and Harold Urey experiments.
- a description of the evolutionary path extraterrestrial life may take and the determination of the number of intelligent extraterrestrial civilizations.
- a description of the plaque and record that were launched with the Pioneer and Voyager spacecraft, respectively.
- a description of Project Ozma, Project SETI, and Project Cyclops.
- a description of the message sent from the 305m Arecibo radio telescope in 1974.

After Viewing the Program

Review the text assignments.

Review the terms that appear in italics or boldfaced type in the text. In particular, note the key terms and concepts listed in the study activities section.

Complete the study activities and the self-test in this study guide; then check your answers with the answer key at the end of the lesson.

Since 1960, several attempts have been made to detect radio signals from distant civilizations. At that time, Project Ozma was the first such attempt made by the United States. One of the most unusual attempts to locate intelligent life via radio signals occurred in 1974: the attempt made by scientists Frank Drake and Carl Sagan. Drake and Sagan sent a message from the 305-meter radio telescope at the Arecibo Observatory, Puerto Rico. The message was transmitted in binary code form toward the globular cluster M13 which is 23,000 light-years from Earth. Since there are several hundred thousand stars in this cluster, it's possible that some civilizations could receive the message. The message contains information about the human body, DNA, a description of our solar system, and a diagram of the Arecibo telescope. Scientists do not expect to receive an answer; the message was sent to indicate our belief that intelligent life can exist somewhere in the Galaxy. Even if a response were sent, we could not expect it until about 50,000 A.D.

```
0 0 0 0 0 0 1 0 1 0 1 0 1 0 0 0 0 0 0 0 0 0 0 0 0 1 0 1 0 0 0 0 0 1 0 1 0
0 0 0 0 0 0 1 0 0 1 0 0 0 1 0 0 0 1 0 0 0 1 0 0 1 0 1 1 0 0 1 0 1 0 1 0 1
0 1 0 1 0 1 0 1 0 1 0 0 1 0 0 1 0 0 0 0 0 0 0 0 0 0 0 0 0 0 0 0 0 0 0 0 0
0 0 0 0 0 0 0 0 0 0 0 0 0 0 0 0 1 1 0 0 0 0 0 0 0 0 0 0 0 0 0 0 0 0 0 0 0
1 1 0 1 0 0 0 0 0 0 0 0 0 0 0 0 0 0 0 0 0 0 0 1 1 0 1 0 0 0 0 0 0 0 0 0 0
0 0 0 0 0 0 0 0 1 0 1 0 1 0 0 0 0 0 0 0 0 0 0 0 0 0 0 0 0 0 0 1 1 1 1 1 0
0 0 0 0 0 0 0 0 0 0 0 0 0 0 0 0 0 0 0 0 0 0 0 0 0 0 0 0 0 0 0 1 1 0 0 0 0
1 1 1 0 0 0 1 1 0 0 0 0 1 1 0 0 0 1 0 0 0 0 0 0 0 0 0 0 0 0 0 1 1 0 0 1 0
0 0 0 1 1 0 1 0 0 0 1 1 0 0 0 1 1 0 0 0 0 1 1 0 1 0 1 1 1 1 1 0 1 1 1 1 1
0 1 1 1 1 1 0 1 1 1 1 1 0 0 0 0 0 0 0 0 0 0 0 0 0 0 0 0 0 0 0 0 0 0 0 0 0
0 1 0 0 0 0 0 0 0 0 0 0 0 0 0 0 0 0 0 1 0 0 0 0 0 0 0 0 0 0 0 0 0 0 0 0 0
0 0 0 0 0 0 0 0 0 0 0 1 0 0 0 0 0 0 0 0 0 0 0 0 0 0 0 0 0 1 1 1 1 1 1 0 0
0 0 0 0 0 0 0 0 0 0 0 1 1 1 1 1 0 0 0 0 0 0 0 0 0 0 0 0 0 0 0 0 0 0 0 0 0
0 0 1 1 0 0 0 0 1 1 0 0 0 0 1 1 1 0 0 0 1 1 0 0 0 1 0 0 0 0 0 0 0 1 0 0 0
0 0 0 0 0 0 1 0 0 0 0 1 1 0 1 0 0 0 0 1 1 0 0 0 1 1 1 0 0 1 1 0 1 0 1 1 1
1 1 0 1 1 1 1 1 0 1 1 1 1 1 0 1 1 1 1 1 0 0 0 0 0 0 0 0 0 0 0 0 0 0 0 0 0
0 0 0 0 0 0 0 0 0 1 0 0 0 0 0 0 1 1 0 0 0 0 0 0 0 0 0 1 0 0 0 0 0 0 0 0 0
0 0 1 1 0 0 0 0 0 0 0 0 0 0 0 0 0 0 0 1 0 0 0 0 0 1 1 0 0 0 0 0 0 0 0 0 0
1 1 1 1 1 1 0 0 0 0 0 1 1 0 0 0 0 0 0 1 1 1 1 1 0 0 0 0 0 0 0 0 0 0 1 1 0
0 0 0 0 0 0 0 0 0 0 0 0 1 0 0 0 0 0 0 0 0 1 0 0 0 0 0 0 0 0 1 0 0 0 0 0 1
0 0 0 0 0 0 1 1 0 0 0 0 0 0 0 1 0 0 0 0 0 0 0 1 1 0 0 0 0 1 1 0 0 0 0 0 0
1 0 0 0 0 0 0 0 0 0 0 1 1 0 0 0 1 0 0 0 0 1 1 0 0 0 0 0 0 0 0 0 0 0 0 0 0
0 1 1 0 0 1 1 0 0 0 0 0 0 0 0 0 0 0 0 0 1 1 0 0 0 1 0 0 0 0 1 1 0 0 0 0 0
0 0 0 0 1 1 0 0 0 0 1 1 0 0 0 0 0 0 1 0 0 0 0 0 0 0 1 0 0 0 0 0 0 1 0 0 0
0 0 0 0 0 1 0 0 0 0 0 1 0 0 0 0 0 0 0 1 1 0 0 0 0 0 0 0 0 1 0 0 0 1 0 0 0
0 0 0 0 0 1 1 0 0 0 0 0 0 0 0 1 0 0 0 1 0 0 0 0 0 0 0 0 0 1 0 0 0 0 0 0 0
1 0 0 0 0 0 1 0 0 0 0 0 0 0 1 0 0 0 0 0 0 0 1 0 0 0 0 0 0 0 1 0 0 0 0 0 0
0 0 0 0 0 0 1 1 0 0 0 0 0 0 0 0 0 1 1 0 0 0 0 0 0 0 0 1 1 0 0 0 0 0 0 0 0
0 1 0 0 0 1 1 1 0 1 0 1 1 0 0 0 0 0 0 0 0 0 0 0 1 0 0 0 0 0 0 0 1 0 0 0 0
0 0 0 0 0 0 0 0 0 0 1 0 0 0 0 0 1 1 1 1 1 0 0 0 0 0 0 0 0 0 0 0 0 1 0 0 0
0 1 0 1 1 1 0 1 0 0 1 0 1 1 0 1 1 0 0 0 0 0 0 1 0 0 1 1 1 0 0 1 0 0 1 1 1
1 1 1 1 0 1 1 1 0 0 0 0 1 1 1 0 0 0 0 0 1 1 0 1 1 1 0 0 0 0 0 0 0 0 0 1 0
1 0 0 0 0 0 1 1 1 0 1 1 0 0 1 0 0 0 0 0 0 1 0 1 0 0 0 0 0 1 1 1 1 1 1 0 0
1 0 0 0 0 0 0 1 0 1 0 0 0 0 0 1 1 0 0 0 0 0 0 1 0 0 0 0 0 1 1 0 1 1 0 0 0
0 0 0 0 0 0 0 0 0 0 0 0 0 0 0 0 0 0 0 0 0 0 0 0 0 0 0 0 0 0 0 0 1 1 1 0 0
0 0 0 1 0 0 0 0 0 0 0 0 0 0 0 0 0 0 1 1 1 0 1 0 1 0 0 0 1 0 1 0 1 0 1 0 1
0 1 0 0 1 1 1 0 0 0 0 0 0 0 0 0 1 0 1 0 1 0 1 0 0 0 0 0 0 0 0 0 0 0 0 0 0
0 0 1 0 1 0 0 0 0 0 0 0 0 0 0 0 0 0 1 1 1 1 1 0 0 0 0 0 0 0 0 0 0 0 0 0 0
0 0 0 1 1 1 1 1 1 1 1 1 0 0 0 0 0 0 0 0 0 0 0 0 1 1 1 0 0 0 0 0 0 0 1 1 1
0 0 0 0 0 0 0 0 0 1 1 0 0 0 0 0 0 0 0 0 0 0 1 1 0 0 0 0 0 0 0 1 1 0 1 0 0
0 0 0 0 0 0 0 1 0 1 1 0 0 0 0 0 1 1 0 0 1 1 0 0 0 0 0 0 0 1 1 0 0 1 1 0 0
0 0 1 0 0 0 1 0 1 0 0 0 0 0 1 0 1 0 0 0 1 0 0 0 0 1 0 0 0 1 0 0 1 0 0 0 1
0 0 1 0 0 0 1 0 0 0 0 0 0 0 0 1 0 0 0 1 0 1 0 0 0 1 0 0 0 0 0 0 0 0 0 0 0
0 1 0 0 0 0 1 0 0 0 0 1 0 0 0 0 0 0 0 0 0 0 0 0 1 0 0 0 0 0 0 0 0 0 1 0 0
0 0 0 0 0 0 0 0 0 0 0 0 1 0 0 1 0 1 0 0 0 0 0 0 0 0 0 0 0 1 1 1 1 0 0 1 1
1 1 1 0 1 0 0 1 1 1 1 0 0 0
```

ARECIBO MESSAGE IN BINARY CODE was transmitted in 1974 toward the Great Cluster in Hercules from the 1,000-foot antenna at Arecibo. The message is decoded by breaking up the characters into 73 consecutive groups of 23 characters each and arranging the groups in sequence one under the other, reading right to left and then top to bottom. The result is a visual message that can be more easily interpreted by making each 0 of binary code represent a white square and each 1 a black square.

Figure 21.1. Arecibo message in binary code.

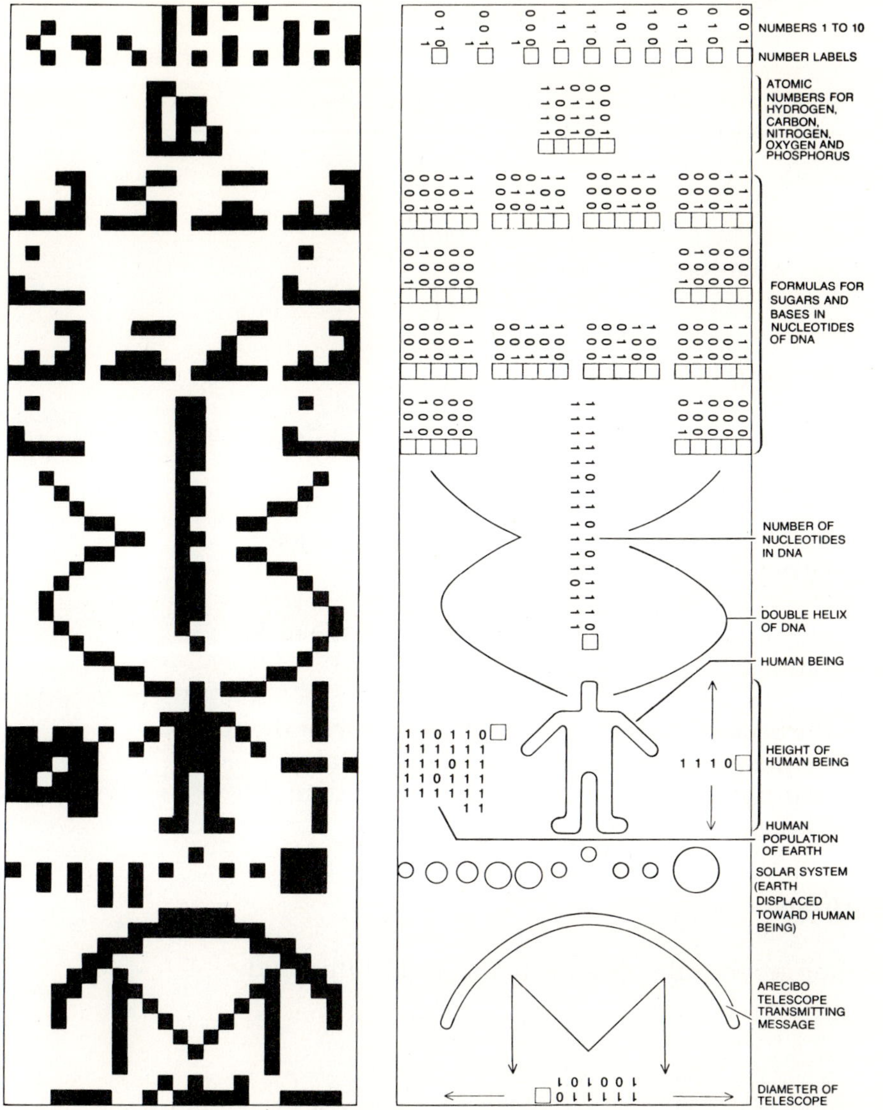

ARECIBO MESSAGE IN PICTURES and accompanying translation shows the binary version of the message decoded. Each number that is used is marked with a label that indicates its start. When all the digits of a number cannot be fitted into one line, the digits for which there is no room are written under the least significant digit. (The message must be oriented in three different ways for all the numbers shown to be read.) The chemical formulas are those for the components of the DNA molecule: the phosphate group, the deoxyribose sugar and the organic bases thymine, adenine, guanine and cytosine. Both the height of the human being and the diameter of the telescope are given in units of the wavelength that is used to transmit the message: 12.6 centimeters.

Figure 21.2. Arecibo message in pictures.

Complete the following exercises: 1, 2, 4, 7, and 8 in text Chapter 28 on page 490.

Study Activities

Key Terms and Concepts

photosynthesis
intelligent life in the Galaxy equation
UFOs
Project Cyclops
Project Ozma
Project SETI

Using What You've Learned

1. What is one serious danger in increasing the carbon dioxide content of the atmosphere?

2. Make a detailed list of the reasons why O, B, A, and M stars are *not* expected to shelter civilizations.

3. Make a detailed list of the reasons why F, G, and K stars are the most likely to shelter civilizations.

4. From Appendix 14 in the back of the text, select the stars nearest the sun most likely to shelter civilizations. Be sure to consider spectral type, luminosity class, and multiplicity (single or multiple star systems). Locate them on the SC1 Star Chart for Lesson Eight.

Recommended Readings

For supplementary reading see the bibliography in the back of this study guide.

Self-Test

(Select the one best answer.)

Objective 1: List the criteria involved in calculating the number of communicative civilizations in the Galaxy.

1. The length of time a civilization continues in its communicative phase is the most difficult factor to estimate accurately.
   a. true
   b. false

Objective 2: Describe the attempts made by astronomers to detect planets in orbit around other stars.

2. Astronomers have conclusively detected planets around other stars. These observations were confirmed by IRAS, the Infrared Astronomy Satellite.
   a. true
   b. false

Objective 3: Describe the modern attempts to contact extraterrestrial civilizations.

3. The Voyager spacecraft carries an interstellar message. It is a (n)
   a. postcard depicting human life on Earth.
   b. audio-video recording of Earth sounds and scenes.
   c. instrument specifically designed to broadcast Earth messages.
   d. television camera showing Earth as it is right now.

4. It is most likely that an Earth-like planet with some form of life on it would be found around an A or B type star.
   a. true
   b. false

5. The first American attempt to detect intelligent signals from extraterrestrial planets was called
   a. Project SETI.
   b. the Very Large Array.
   c. Project Ozma.
   d. Project Life Search.

6. A proposed network of more than 1,000 linked radio telescopes designed to detect extraterrestrial radio sources was called
   a. Project SETI.
   b. the Very Large Array.
   c. Arecibo.
   d. Project Cyclops.

7. Astronomers feel that the wavelength region around 21 centimeters would be most likely selected by an extraterrestrial intelligence to broadcast messages.
   a. true
   b. false

8. A globular cluster was the target for the
   a. Voyager spacecraft.
   b. Pioneer spacecraft.
   c. first American message sent from Earth into outer space.
   d. Mariner probes.

Objective 4: Describe the conditions necessary for a solar system to contain planets with the possibility of life evolving on them.

9. We have not been able to detect any organic compounds in gas clouds far out in space.
   a. true
   b. false

10. It is highly probable that our sun is not the only star with planets in orbit around it.
    a. true
    b. false

Answer Key

1. a	6. d
2. b	7. a
3. b	8. c
4. b	9. b
5. c	10. a

Relativity 22

Overview

Before Galileo's time, common sense held that a heavier object falls faster than a light one, but a few experiments by this scientist and thinker demolished that bit of common sense. Newton's laws of gravitation explained how, allowing for air resistance, the acceleration of gravity was caused by the combined masses of the attracting bodies, not by the weight of one of them. Three or four generations have grown up in what we might call the "age of relativity," so perhaps some of the basic principles of relativity are a little less jarring to common sense today than they were in the early years of this century. Despite this increased familiarity, the principles of relativity and the phenomena that the theories describe are quite removed from everyday experience, as you'll discover in this lesson. If so, then why study relativity?

We study relativity because Einstein's theories explain some aspects of motion and gravity that Newton's laws couldn't explain. They also make it possible to explain the massive production of energy within the stars and provide a basis for predicting the behavior of particles escaping from the stars. These same theories can be used to predict many of the phenomena you'll study in the remaining lessons, such as the black holes formed by massive collapsed stars, and to develop a model of the formation of the universe from a single unimaginable explosion. In short, the special and general theories of relativity have sparked a revolution in astronomy and many other sciences.

The most precise and elegant description of relativity is found in the language of mathematics. Yet it's entirely possible to gain a good understanding of relativity by studying examples, both actual and hypothetical, of events that take place in the universe. What you have to do is readjust your thinking just as you did when you started thinking in terms of the solar system and galactic distances instead of earthbound measurements. For example: The inertia (or mass) of an object is not absolute; the flow of time is not absolute; matter and energy are not absolute properties, because one can be changed into the other. The speed of light, however, is absolute; it stays the same regardless of any movement of the source or the observer. And there are a few more examples. In this lesson you'll read about as well as see examples that will make these rather strange-seeming behaviors of matter, time, and space understandable and better appreciated.

If this is your first encounter with Einstein's theories, they may seem a bit "disorienting." Your feelings may resemble those of the people who had to trade the Ptolemaic solid Earth for the Copernican Earth-in-motion. Neither an orbiting Earth nor the relativistic universe is any less stable than what we imagined them to be before. What each discovery gave to humankind, instead, was a new and better way of looking at and understanding the universe.

Learning Objectives

When you have completed all assignments in this lesson, you should be able to:

1. State the basic postulates of the special and general theories of relativity.

2. State the major predictions of the special and general theories of relativity and list their observational verifications.

Assignments for This Lesson

Before Viewing the Program

Read the overview at the beginning of this lesson and look over the learning objectives that precede the assignments.

Read the introduction to Part IV of the text, pages 332-334, and all of Chapters 19 and 20.

View Program 22, "Relativity"

As you view the program, look for:

- a description of the determination of the speed of light.
- the Michelson-Morley experiment with the interferometer.
- the principles of special relativity and the results of experiments verifying those principles.
- an explanation of the "twin paradox."
- the principle of equivalence and the development of the general theory of relativity; and a description of the experiments verifying the general theory of relativity.
- a description of the four-dimensional nature of spacetime and the meaning of the curvature of spacetime.

After Viewing the Program

Review the text assignments.

Review the terms that appear in italics or boldfaced type in the text. In particular, note the key terms and concepts listed in the study activities section.

Complete the study activities and the self-test in this study guide; then check your answers with the answer key at the end of the lesson.

Complete the following exercises: 5, 6 and 7 in text Chapter 19, page 347, and exercises 2 and 4 in text Chapter 20, page 360.

Study Activities

Key Terms and Concepts

Michelson-Morley experiment
speed of light (absoluteness)
principle of relativity
time dilation
length contraction
mass increase with velocity
$E = mc^2$
tachyons
spacetime
principles of equivalence
gravitational redshift
curvature of spacetime
gravitational waves

Using What You've Learned

For questions 1 through 4 choose among the following:

a geodesic
b. deflection of light grazing the sun
c. gravitational redshift
d. gravitational waves

1. Which is (or has been) observed in the light from dwarf stars, in gamma rays in the Jefferson Laboratory experiment, and in radiation from a hydrogen maser on the ground as observed from a rock 10,000 km above Earth's surface?

2. Which is the path of a body in spacetime?

3. Which is best observed during a solar eclipse?

4. Which is produced by a redistribution of matter?

5. Read more about the faster-than-light particle called the tachyon. A good source is the book *The Cosmic Frontiers of General Relativity* by William J. Kaufmann III (Boston: Little Brown, 1977), pages 42-45. Report on what you have learned.

Recommended Readings

For supplementary reading see the bibliography in the back of this study guide.

Self-Test

(Select the one best answer.)

Objective 1: State the basic postulates of the special and general theories of relativity.

1. The speed of light
   a. depends on the relative motion of the observers.
   b. is the same for all observers.
   c. depends on the position of the source.
   d. depends on the relative motion between the sources and the observer.

2. The speed of light is
   a. about 30 km/sec.
   b. about 3,000 km/sec.
   c. about 300,000 km/sec.
   d. infinite.

3. If one spaceship is moving at .9c (nine-tenths the speed of light) away from Earth and another spaceship is moving away from Earth in the opposite direction at the same speed, what would be the relative velocity between the two spaceships?
   a. some speed between .9c and c
   b. .9c
   c. some speed between c and 1.8c
   d. exactly 1.8c

4. Joe is sent on a trip in a spaceship and travels around for a while at near light speeds. His twin brother Pat remains on Earth. When Joe returns he would be
   a. the same age as his brother.
   b. older than Pat.
   c. younger than Pat.

5. It is impossible to distinguish between a smooth acceleration and being motionless in a gravitational field. This is a statement of
   a. the principle of equivalence.
   b. the constant velocity of light principle.
   c. the curvature of spacetime principle.
   d. the principle of the consistency in the laws of physics.

6. In a gravitational field greater than Earth's, time would
   a. speed up.
   b. slow down.
   c. remain the same.

Objective 2: State the major predictions of the special and general theories of relativity and list their observational verifications.

7. Because of the near light speeds of the muon, Earth-based observers see the lifetime of the muon extended. Time slows for the muon. If we could ride on the muon as it approached Earth, we would see
   a. time speed up for the muon.
   b. time slow down for the muon.
   c. the muon never reach the ground.
   d. the distance between the muon and the ground to be shorter than that same distance measured by an Earth-based observer.

8. The light from a distant star that passes near the edge of the sun would cause the position of the star to
   a. appear farther from the sun when compared to its true position in the sky.
   b. appear closer to the sun when compared to its true position in the sky.
   c. be unchanged.

9. The precession of the orbit of Mercury is explained by taking into account the gravitational effects of the other eight planets and the relativistic effects due to the sun's gravity.
   a. true
   b. false

10. Of the following, which are effects predicted by the general theory of relativity?
   a. shift in spectral lines due to velocity
   b. splitting of spectral lines due to magnetism
   c. constant speed of light
   d. gravity waves

Answer Key

1. b	6. b
2. c	7. d
3. a	8. a
4. c	9. a
5. a	10. d

Stars — The Nuclear Furnace 23

Overview

In Lesson Fifteen you took a brief look into the interior of the sun. You'll soon discover, by looking at the first few pages of your reading assignment, that this lesson offers a much more penetrating look into the interiors of suns. In Lesson Fifteen you were introduced to Einstein's formula, $E = mc^2$, the formula that explains the conversion of matter into energy. The lesson you just studied on relativity explains Einstein's formula in much more detail. It's a good idea to keep this formula firmly in mind as you study this lesson on stars.

Barring only the tenuous nebulas and the galaxies themselves, the stars are the largest structures in our universe. It may seem paradoxical that the key to understanding stellar processes lies in the smallest discrete object in the universe, the atom. Until this century, atoms were thought to be the smallest units of matter that exist. While this is, in a sense, still true, physicists have penetrated the atom itself to learn about the "building blocks" that form each one. In this lesson, you'll be introduced to "hadrons" and "quarks" — inhabitants of the most minute frontiers yet investigated by science.

As you fit together these pieces of current scientific understanding, you can construct your own mental picture of a star's interior, much like astronomers who carefully construct a complete and detailed picture called a stellar model. Both the text and television program describe the detail and complexity involved in constructing a model of a star. The current stellar models give us probable answers to questions that, until recently, would have been almost impossible to speculate on: Which of the stars we see are youngest? Which are oldest? What determines the colors of stars? What's the expected life span of our sun?

This lesson also describes the processes by which a star is formed out of the dust and gases of interstellar space. These are events which would be invisible to our best telescopes, even if we had the long years of observation time that would be required to watch them progress. However, infrared detectors and radio telescopes do offer astronomers new opportunities to observe some of these processes firsthand.

Learning Objectives

When you have completed all assignments in this lesson, you should be able to:

1. Describe the basic equilibrium conditions involved in stellar models.
2. List and describe the steps in constructing a stellar model.
3. List and describe the observational data which must be known to predict the evolution of a star.
4. Describe the relationship between rate of energy production and stellar lifetime.
5. List the different types of nuclear reactions which are known to occur in stellar interiors.
6. List and describe the gross properties of a protostar.
7. Explain the relationship between the lifetime of a protostar and its mass.
8. List chronologically and describe the major steps in the evolution of a protostar into a zero-age main sequence star.
9. Describe the nature and consequences of antimatter.

Assignments for This Lesson

Before Viewing the Program

Read the overview at the beginning of this lesson and look over the learning objectives that precede the assignments.

Reread all of text Chapter 21 and pages 372-376 of Chapter 22.

View Program 23, "Stars – The Nuclear Furnace"

As you view the program, look for:

- an explanation of why stars shine and the nature of the nuclear force.
- the proton-proton chain reaction of nuclear fusion.
- an explanation of the relationship between the mass of a star and its nuclear fusion rate, temperature, size, and lifetime.
- the triple-alpha process of nuclear fusion.
- the formation of stars from nebulas and the observational evidence supporting this formation.
- an explanation of how stars would possibly form from a supernova explosion.

After Viewing the Program

Review the text assignments.

Review the terms that appear in italics or boldfaced type in the text. In particular, note the key terms and concepts listed in the study activities section.

Complete the study activities and the self-test in this study guide; then check your answers with the answer key at the end of the lesson.

Complete the following exercises: 3, 6, and 8 in text Chapter 21, page 371, and exercises 1 and 3 in text Chapter 22, page 387.

Study Activities

Key Terms and Concepts

hydrostatic equilibrium
thermal equilibrium
nucleus (of the atom)
quarks
antimatter
fusion
fission
zero-age main sequence
protostar

Using What You've Learned

1. From other sources, acquire a list of the characteristics of T Tauri stars, Herbig-Haro objects, and Bok globules. Compare this information with what you have already learned about the characteristics of protostars.

2. List the conditions used in constructing a model for a star.

3. Select a particle found in the nucleus of the atom (there are over 300 particles) such as a proton, neutron, meson, or an antiparticle such as a positron or an antiproton. Write a report on the particle's characteristics, the history of its discovery, and what effects, if any, it has had on our understanding of the nature of matter. Be sure to list your references.

Recommended Readings

For supplementary reading see the bibliography in the back of this study guide.

Self-Test

(Select the one best answer.)

Objective 1: Describe the basic equilibrium conditions involved in stellar models.

1. Conduction is the least important form of energy transportation in a main sequence star.
   a. true
   b. false

2. A balance between the amount of energy produced in the core of the star and the amount of energy released at the surface is called
   a. hydrostatic equilibrium.
   b. thermal equilibrium.
   c. luminosity equilibrium.
   d. energy-mass equilibrium.

Objective 2: List and describe the steps in constructing a stellar model.

3. Gamma rays are produced in the ______________ of main sequence stars.
   a. atmospheres
   b. cores
   c. coronas
   d. photospheres

4. Which of the following must be employed to develop a model of the interior of a star?
   a. a set of physical assumptions
   b. mathematical equations
   c. the perfect gas law
   d. all of the above

Objective 2 and Objective 3: List and describe the observational data which must be known to predict the evolution of a star.

5. The development of a stellar model does not need to be consistent with the observed activity of a star.
   a. true
   b. false

6. The formation of stars may, in some cases, be triggered by a supernova event.
   a. true
   b. false

Objective 4: Describe the relationship between rate of energy production and stellar lifetime.

7. A K type star will spend more time on the main sequence of the H-R diagram than what other type (s) of stars?
   a. A
   b. B
   c. G
   d. all of the above

Objective 5: List the different types of nuclear reactions which are known to occur in stellar interiors.

8. Which of the following is the most massive?
   a. proton
   b. electron
   c. positron
   d. helium nucleus

9. The most energetic form of nuclear energy is
   a. nuclear fusion.
   b. nuclear fission
   c. matter-antimatter annihilation.

10. Nuclear fusion of hydrogen into helium occurs in the cores of
    a. protostars.
    b. red giants.
    c. main sequence stars. (IX)

Objective 6: List and describe the gross properties of a protostar.

11. Gravitational contraction is the primary source of a star's energy during its
    a. main sequence life.
    b. early stages as a protostar.
    c. red giant life.

12. What wavelength of radiation would be detected first from a protostar?
    a. visible
    b. infrared
    c. ultraviolet
    d. X-ray

Objective 7: Explain the relationship between the lifetime of a protostar and its mass.

13. A globule would appear as a
    a. dark, round blob seen silhouetted against a brighter background.
    b. bright spherical object in outer space.
    c. bright cloud of gas in a star-forming region.
    d. dark distended cloud against a bright background.

Objective 8: List chronologically and describe the major steps in the evolution of a protostar into a zero-age main sequence star.

14. When nuclear fusion reactions begin to supply all energy, and the star stops contracting, the star is said to be
    a. a protostar.
    b. a nova.
    c. on the zero-age main sequence.
    d. emitting visible light only.

15. O type stars would spend less time in the protostar phase than what other types of stars?
    a. B
    b. G
    c. K
    d. all of the above

16. Mass is the most important characteristic in determining the lifetime of a protostar and main sequence star.
    a. true
    b. false

Objective 9: Describe the nature and consequences of antimatter.

17. The balance between the outward gas and radiation pressure and the inward gravity is called ______________ equilibrium.
    a. thermal
    b. photospheric
    c. hydrostatic
    d. nuclear synthesis

18. The antimatter counterpart to an electron is the
    a. positron.
    b. meson.
    c. neutrino.
    d. antiproton.

Answer Key

1. a	7. d	13. a
2. b	8. d	14. c
3. b	9. c	15. d
4. d	10. c	16. a
5. b	11. b	17. c
6. a	12. b	18. a

White Dwarfs and Red Giants 24

Overview

Has it occurred to you how nice it would be to have scientifically based records of observation stretching back perhaps 10,000 years or even half a million years? All the years since Galileo first put his eye to a telescope amount to less than an instant in the life of the Milky Way or even our own sun. The entire period covered by human history since the beginning of written language isn't long enough for even the most energetic and largest star to have evolved from its early protostar stage to zero-age main sequence. For all of its effort, astronomy's role might be compared to that of a scientist who had to construct his theories based on only one snapshot of a complex phenomenon. Of course, the situation for astronomers is not quite that desperate. In this lesson you'll learn about a method used to compare stars at different stages in their evolution. It's as if the single "snapshot" contained a tremendous amount of data about virtually every part of the complex phenomenon.

In the last lesson you learned about the birth of stars and how our own sun came into being some five thousand million years ago. Fortunately, our sun is destined for a long, long life as a main sequence star, at least according to generally accepted present-day theories. But like all stars, the sun has a finite life span, which will terminate with the sun reduced to a small dim white dwarf about the size of Earth. Before that happens, though, the sun will undergo some spectacular changes that can be followed best by referring to the H-R diagram.

In this lesson, you'll learn there are only three possible conclusions to a star's life, one of which is the white dwarf. The existence of variable stars is explained in some detail, and three types of "dependable" pulsating stars (those whose period of variation helps to determine their absolute magnitude) are illustrated in the television program. As the title of the lesson suggests, you'll also learn about the "red giant" stage of a star's life. Following the red giant phase, some stars evolve into planetary nebulas. A planetary nebula is a rather short-lived space "bubble" that is related to the death of stars. Finally, this lesson discusses why some stars become novas, temporarily and rapidly increasing in brightness only to fade again.

Learning Objectives

When you have completed all assignments in this lesson, you should be able to:

1. Describe the relations involved in main sequence turn-off points.
2. Describe the Chandrasekhar limit.
3. Describe the process involved in stellar giant and supergiant formation.
4. List and describe the steps in the formation of planetary nebulas.
5. List and describe the steps in the formation of white dwarf stars.
6. Determine the relative ages of star clusters.
7. Explain the correlation between stellar cluster types and ages.
8. List and describe the types of pulsating variables.
9. Describe the pulsation mechanism.
10. List and describe the observed properties of eruptive variables other than supernovas.

Assignments for This Lesson

Before Viewing the Program

Read the overview at the beginning of this lesson and look over the learning objectives that precede the assignments.

Read the section "Evolution from the Main Sequence to Giants" in text Chapter 22, beginning on page 376; then read to the end of the chapter.

Read the following sections in text Chapter 23: "White Dwarfs" and "Evolution of the Sun," pages 388-394. Reread the sections of Chapter 15 on cepheid and RR Lyrae variable stars, pages 274-275.

View Program 24, "White Dwarfs and Red Giants"

As you view the program, look for:

- the description of the evolution of our sun and the effects on Earth.
- the description of the formation of a red giant star.
- the description of the behavior of different types of variable stars (RR Lyrae, cepheids, Miras) and the explanation of the period-luminosity relation.
- the description of the formation of white dwarfs and planetary nebulas.
- the explanation of the event called a nova (not to be confused with supernova).

After Viewing the Program

Review the text assignments.

Review the terms that appear in italics or boldfaced type in the text. In particular, note the key terms and concepts listed in the study activities section.

Complete the study activities and the self-test in this study guide; then check your answers with the answer key at the end of the lesson.

Complete the following exercises: 2, 5, and 7 in text Chapter 22, page 387, and exercises 2 and 3 in text Chapter 23, pages 406-407.

Study Activities

Key Terms and Concepts

electron degeneracy
planetary nebula
nova
Problicom (Ben Mayer)
Chandrasekhar limit
helium flash

Using What You've Learned

1. Using the H-R diagram you constructed in Lesson Seventeen or sketching a new one (follow the example on page 293 in your text, Figure 16.9), locate each of the following stars on the diagram.
   a. a star converting hydrogen to helium in its core
   b. a star converting hydrogen to helium in an envelope surrounding the core
   c. a massive, young star
   d. a star pulsating with a regular period
   e a star converting helium to carbon in the core
   f. a very old and dense star
   g. a protostar

2. Arrange these clusters in order of increasing age.

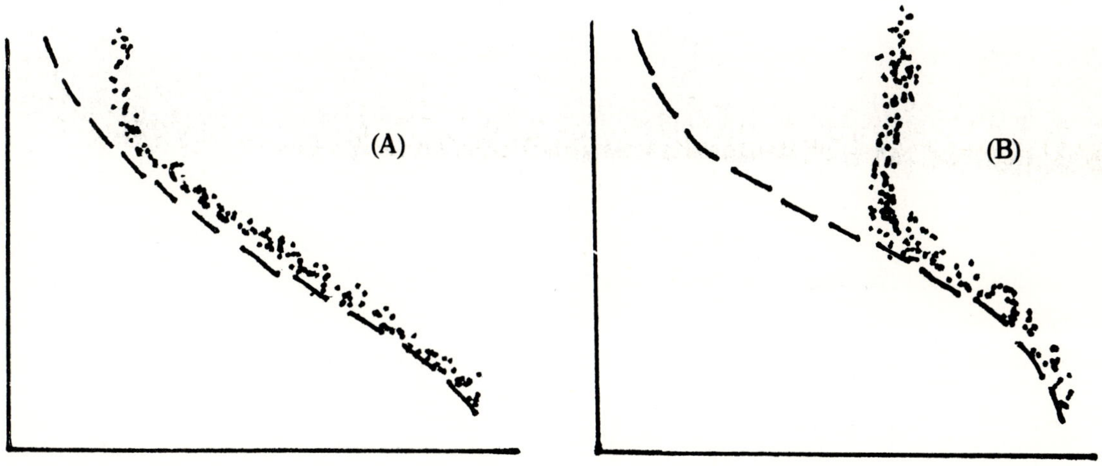

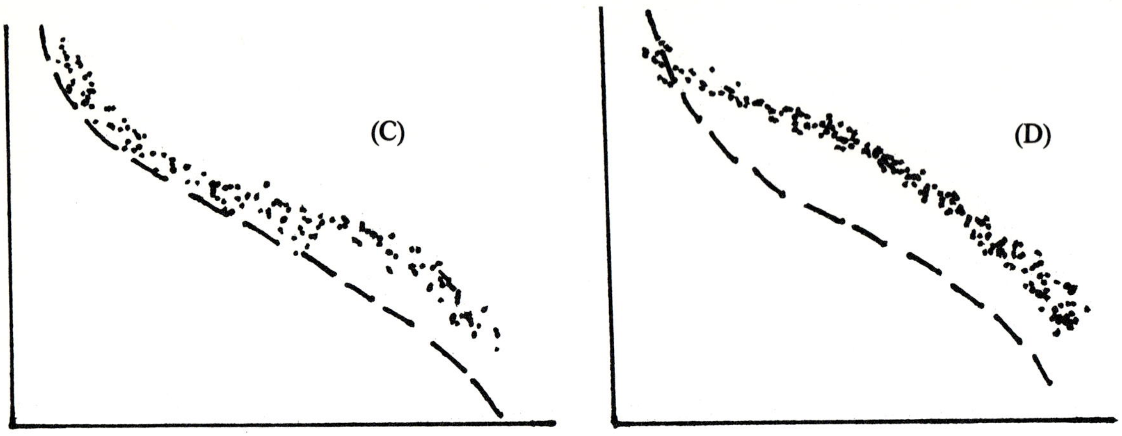

Figure 24.1. Star clusters.

Recommended Readings

For supplementary reading see the bibliography in the back of this study guide.

Self-Test

(Select the one best answer.)

Objective 1: Describe the relations involved in main sequence turn-off points.

1. Constructing an H-R diagram of a star cluster and determining the "turn-off" point from the main sequence helps to determine the cluster's
   a. mass.
   b. size.
   c. distance from the center of the Galaxy.
   d. age.

Objective 2: Describe the Chandrasekhar limit.

2. 1.4 solar masses (the Chandrasekhar limit) is the
   a. maximum mass for a white dwarf.
   b. maximum mass for a red giant.
   c. minimum mass for a star to initiate hydrogen fusion.
   d. minimum mass for a star to initiate helium fusion.

Objective 3: Describe the process involved in stellar giant and supergiant formation.

3. The nuclear fusion of helium into carbon occurs at 100 million °K in
   a. main sequence stars.
   b. blue giants.
   c. the cores of red giants.
   d. the hydrogen envelope around white dwarfs.

Objective 4: List and describe the steps in the formation of planetary nebulas.

4. The relatively slow ejection of a red giant's outer envelope of gas describes
   a. a planetary nebula.
   b. the formation of a white dwarf.
   c. a nova.
   d. the evolution of a main sequence star.

Objective 5: List and describe the steps in the formation of white dwarf stars.

5. The light and energy emitted by a white dwarf are produced by thermonuclear reactions.
   a. true
   b. false

6. Our sun will, to the best of our knowledge, end its life as a
   a. white dwarf, cooling off and going out.
   b. nova.
   c. red giant.
   d. T Tauri star.

Objective 6: Determine the relative ages of star clusters.

7. Given the spectral types of star clusters at which they "turn off" the main sequence, which cluster is the youngest?
   a. M2
   b. G6
   c. B4
   d. F0

Objective 7: Explain the correlation between stellar cluster types and ages.

8. Stars that are older and weaker in metals are more frequently found in
   a. open clusters.
   b. globular clusters.

Objective 8: List and describe the types of pulsating variables.

9. The plot of a cepheid variable on the H-R diagram would most likely find it in the
   a. red giant branch.
   b. main sequence.
   c. white dwarf branch.
   d. Hertzsprung gap.

Objective 9: Describe the pulsation mechanism.

10. The variation in brightness of a cepheid occurs because it is
    a. being eclipsed by its binary companion.
    b. exploding.
    c. changing in diameter in a very regular manner.
    d. fluctuating violently in size.

Objective 10: List and describe the observed properties of eruptive variables other than supernovas.

11. The nova outburst is believed to involve a
    a. single star exploding.
    b. binary star system with one of the stars being a white dwarf.
    c. red dwarf.
    d. cepheid variable.

Answer Key

1. d
2. a
3. c
4. a
5. b
6. a
7. c
8. b
9. d
10. c
11. b

Supernovas and Pulsars 25

Overview

Two of the brightest and most distinctive stars of Earth's night sky will give us a fantastic fireworks show sometime in the next 10,000 years. Science cannot be any more precise in pinpointing the time, because there's no way to tell just when interior conditions of either star will reach a critical point. Betelgeuse, some 150 parsecs distant, shines brighter than magnitude 1 in the shoulder of Orion. Antares is located in the heart of Scorpio. In a dark sky, you probably can detect the slight orange color of these stars, for both are tremendous red supergiants that are much more massive than the sun, too massive to take the "peaceful" route to the white dwarf state. The end awaiting them is what Plavec described as "violent" in the television program for Lesson Twenty-four: They will become supernovas (or, in Latin, supernovae).

Comparing a nova to a supernova is something like comparing a firecracker to a volcano. Supernovas are such tremendous stellar explosions that they may shine with light greater than an entire galaxy, maintaining visibility even in Earth's daylight. Supernovas are believed to be the source of all the heavier elements formed in the Galaxy, a trigger for the formation of some new stars from interstellar dust and gas, and a major source of cosmic rays.

In this lesson, you'll learn about the events that lead to the supernova explosion, something of the composition of the material blown outward from the explosion, and about the formation of a neutron star, which inevitably follows if the matter remaining and compressed by the explosion falls within certain mass limits. You'll learn about the properties of neutron stars. Neutron stars contain degenerate matter packed much more densely than in white dwarfs. These stars rotate very rapidly, and they have stronger gravitational and magnetic fields than normal stars. These properties, in fact, led to the discovery and identification of the first neutron star in 1967, in the form of a "pulsar." The rapidly rotating star emits a beam of radiation, something like a searchlight beam, which appears as a series of "flashes" (at the rate of many times each second) to an observer on Earth.

Neutron stars were long known in theory, but were not discovered until the late 1960s. The energy output of most pulsars or neutron stars is strongest in the radio portion of the spectrum, although several visual pulsars and X-ray pulsars have been detected. A fairly recent and significant discovery, which is described in the television program, is a mysterious object named SS 433. This object seems to be ejecting matter in a "searchlight beam" at a speed one-quarter the speed of light. Although not fully understood as yet, this unusual behavior seems to be linked to a neutron star.

Learning Objectives

When you have completed all assignments in this lesson, you should be able to:

1. List and describe the conditions leading to the formation of supernovas.
2. Describe the characteristics and properties of pulsars and neutron stars.
3. Describe the steps in the formation of pulsars and neutron stars.
4. List and describe the problems in observing pulsars and neutron stars.
5. Describe the nature and origin of cosmic rays.

Assignments for This Lesson

Before Viewing the Program

Read the overview at the beginning of this lesson and look over the learning objectives that precede the assignments.

Read the sections "The Evolution of Massive Stars," in text Chapter 23, pages 394-400.

View Program 25, "Supernovas and Pulsars"

As you view the program, look for:

- a description of the supernova seen in 1054 A.D., the remnant of which is called the Crab nebula.
- an explanation of the events leading to a supernova and the formation of a neutron star.
- a description of the discovery of the first pulsar and how the pulsar emits its energy.
- a description of the behavior of the Crab nebula pulsar.
- a description of the formation and observation of cosmic rays.

After Viewing the Program

Review the text assignments.

Review the terms that appear in italics or boldfaced type in the text. In particular, note the key terms and concepts listed in the study activities section.

Complete the study activities and the self-test in this study guide; then check your answers with the answer key at the end of the lesson.

Complete the following exercises: 5, 6, and 7 in text Chapter 23, page 407.

Study Activities

Key Terms and Concepts

pulsars
neutron stars
supernovas (supernovae)
nucleosynthesis
cosmic ray

Using What You've Learned

1. Arrange the following phases of stellar evolution from earliest to latest.
   a. red giant
   b. condensation
   c. main sequence
   d. white dwarf
   e. RR Lyrae star

2. Compare the differences between a nova and a supernova.

3. Investigate how pulsars were first discovered and the steps that led to the conclusion that these objects were neutron stars.

Recommended Readings

For supplementary reading see the bibliography in the back of this study guide.

Self-Test

(Select the one best answer.)

Objective 1: List and describe the conditions leading to the formation of supernovas.

1. Supernova explosions occur in stars that are
   a. white dwarfs.
   b. less massive than our sun.
   c. more massive than our sun.
   d. red dwarfs.

2. The element iron is unique because it
   a. helps to continue the fusion process beyond the element carbon.
   b. cannot undergo any further thermonuclear fusion reactions and generate more energy.
   c. is heavier than carbon.
   d. burns hotter than helium.

3. How many times can a star supernova?
   a. 1
   b. 2
   c. 10
   d. an undetermined number

4. The Crab nebula is a (n)
   a. planetary nebula.
   b. HI region.
   c. galaxy.
   d. supernova remnant.

Objective 2: Describe the characteristics and properties of pulsars and neutron stars.

5. Radio waves are the primary wavelengths emitted by the pulses of a pulsar.
   a. true
   b. false

6. A rapidly spinning sphere of degenerate neutrons best describes a
   a. supernova remnant.
   b. planetary nebula.
   c. pulsar.
   d. white dwarf.

7. Neutron stars are typically
   a. the size of Earth.
   b. 10 miles in diameter.
   c. the size of the moon.
   d. the size of Jupiter.

8. The fastest known pulsar pulsates at
   a. 3 times per second.
   b. 30 times per second.
   c. 60 times per second.
   d. 640 times per second.

9. As a pulsar ages, it
   a. speeds up.
   b. slows down.
   c. remains unchanged.

Objective 3: Describe the steps in the formation of pulsars and neutron stars.

10. A pulsar is formed when
    a. a white dwarf cools and goes out.
    b. a white dwarf begins spinning rapidly after the supernova event.
    c. protons and electrons are forced together to form neutrons as a result of a supernova event.
    d. interstellar gas clouds compress to extremely high densities.

Objective 4: List and describe the problems in observing pulsars and neutron stars.

11. Because we detect signals from pulsars whose beam of energy is directed toward Earth, it is highly likely that there are pulsars in our Galaxy that we have not yet detected and probably never will.
    a. true
    b. false

Objective 5: Describe the nature and origin of cosmic rays.

12. It is believed that neutron stars may produce cosmic rays. Cosmic rays are
    a. particles of matters such as protons and helium nuclei.
    b. high energy electromagnetic radiation.
    c. low energy electromagnetic radiation.
    d. primarily gamma and X-rays.

Answer Key

1. c	7. b
2. b	8. d
3. a	9. b
4. d	10. c
5. a	11. a
6. c	12. a

Black Holes 26

Overview

In the late 1970s and early 1980s, one of the most popular subjects in the media came from astronomy: black holes. At least one elaborately produced science-fiction motion picture, several popularized science books and articles, and a few television documentaries took as their theme these strange distortions of normal matter, space, and time. Some creative thinkers, who did not base their thoughts on the general theory of relavity, speculated that these black holes might be a gateway into another dimension or another universe. Black holes were sometimes pictured as "supergravity centers" that could swallow entire galaxies. Even if you take a more rational approach, visualizing a black hole in space can still leave you a bit awe-struck. To be sure, no one has actually observed a black hole (in fact, by definition, a black hole can't be seen). Both the authors of the text and Edwin Krupp in the television program state that black holes are, as of this date, based on theory, much as neutron stars were until 1967. However, there are observations indicating very strongly that these objects do exist.

The concept itself is really quite simple. You'll recall that an early test of the general theory of relativity demonstrated that gravity bends light. What would happen, then, if the strength of a gravitational field were increased? At some point, light would be bent so much that it would orbit the object. Increasing the gravitational pull even more would draw the light toward the center of the mass. Then no light (or any other form of electromagnetic radiation) could escape.

In this lesson, you'll quickly review some of the information on the death of stars that was presented in the last two lessons. In particular, you'll be reminded that there is a limit to the mass of a star that can become a white dwarf and a similar limit to the mass of a neutron star. It's always possible that a star may expel so much of its mass in an explosion that the remaining matter will become a white dwarf or neutron star. This lesson explores, in theory, what can occur when material exceeding the maximum limits for a white dwarf and neutron star collapses. In the television program you'll observe the structure (as much as is known) of a black hole and learn how such a distortion or warping of space would affect matter around it. You'll learn why X-ray astronomy is employed now in search for black holes and why detection of gravity waves could also aid this search.

Learning Objectives

When you have completed all assignments in this lesson, you should be able to:

1. Describe black holcs and how they behave.

2. Describe the probable steps in the formation of black holes.

3. List and explain some of the problems in observing black holes.

4. Summarize the life cycles of stars by outlining chronologically the stages from birth to death of stars with masses less than, equal to, and greater than that of the sun.

5. Describe the detection of gravity waves.

Assignments for This Lesson

Before Viewing the Program

Read the overview at the beginning of this lesson and look over the learning objectives that precede the assignments.

Read in text Chapter 23 the section "Black Holes," pages 400-406.

View Program 26, "Black Holes"

As you view the program, look for:

- the description of the formation of a black hole.
- the explanation of the relationship between the formation of a black hole and general relativity.
- the description of the various parts of a black hole (photon sphere, event horizon, singularity).
- the explanation of the observations needed to detect a black hole.
- evidence indicating that Cygnus X-1 is a black hole.
- the description of the nature and detection of gravity waves.

After Viewing the Program

Review the text assignments.

Review the terms that appear in italics or boldfaced type in the text. In particular, note the key terms and concepts listed in the study activities section.

Complete the study activities and the self-test in this study guide; then check your answers with the answer key at the end of the lesson.

Complete the following exercises: 7, 8, 9, and 10 in text Chapter 23, page 407.

Study Activities

Key Terms and Concepts

black hole
singularity
event horizon (Schwarzchild radius)
photon sphere
accretion disk
Cygnus X-1
LMC X-3
gravitational redshift
pair production

Using What You've Learned

If you are adventurous, you may wish to do some additional reading and write a report about the theoretical structure of black holes. Your report should include the nature of spacetime within the black hole, a spacetime diagram of a black hole, and a hypothetical journey through a black hole.

Nonmathematical descriptions of these topics are well presented in two books by William J. Kaufmann III. The titles are *Relativity and Cosmology*, 2nd edition (New York: Harper & Row, 1977), and *The Cosmic Frontiers of General Relativity* (Boston: Little, Brown, 1977).

Recommended Readings

For supplementary reading see the bibliography in the back of this study guide.

Self-Test

(Select the one best answer.)

Objective 1: Describe black holes and how they behave.

1. The circular orbit of a photon around a black hole is called the
   a. singularity.
   b. photon sphere.
   c. event horizon.
   d. orbit of light.

2. The gravitational influence of a one solar mass black hole at a distance of 100,000,000 miles would be ____________ the sun's influence.
   a. many times stronger than
   b. many times weaker than
   c. about the same as

Objective 2: Describe the probable steps in the formation of black holes.

3. Which of the following do scientists believe are black holes?
   a. Crab nebula
   b. Cygnus X-1
   c. ring nebula
   d. LMC X-3

4. Three (3) solar masses is the minimum mass an object must have in order to
   a. form a neutron star.
   b. form a white dwarf.
   c. collapse into a black hole.
   d. evolve into a red giant.

Objective 3: List and explain some of the problems in observing black holes.

5. In order to detect a black hole, the object must be associated with another star, so that its gravitational effects can be observed.
   a. true
   b. false

6. Astronomers have been able to detect X-rays that they believe are produced by black holes. If this is the case, these X-rays originate from
   a. the black hole itself.
   b. a nearby companion star.
   c. the photon sphere.
   d. the accretion disk.

Objective 4: Summarize the life cycles of stars by outlining chronologically the stages from birth to death of stars with masses less than, equal to, and greater than that of the sun.

7. Most of a star's lifetime is spent as a
   a. main sequence star.
   b. red giant.
   c. protostar.
   d. nova.

8. What is the present evolutionary stage of most of the red dwarfs (M type-main sequence) that have ever formed in our Galaxy? They are
   a. still main sequence stars.
   b. red giants.
   c. white dwarfs.
   d. variables.

Objective 5: Describe the detection of gravity waves.

9. The most important property of a star that determines its evolutionary sequence is its
   a. temperature.
   b. size.
   c. mass.
   d. chemical composition.

10. During the formation of which of the following objects would gravity waves most likely to be detected? (This would only be true if we actually had operational gravity telescopes.)
    a. planets
    b. solar mass stars
    c. red giants
    d. black holes

Answer Key

1. b	6. d
2. c	7. a
3. b, d	8. a
4. c	9. c
5. a	10. d

Galaxies 27

Overview

Astronomy again stands at a threshold much like that which opened with the invention of the optical telescope. Refinements in photography, improvements in optics, and new telescopes which probe the nonvisible portions of the electromagnetic spectrum have made it possible to see into the vast reaches of spacetime that lie beyond our Galaxy. As astronomers study events and objects lying billions of light-years away, they are looking back in time at the radiations that these objects made billions of years ago.

This lesson and those that follow take you beyond the stars, in a very real sense. Philosophers had often questioned whether "island universes" existed beyond the star-filled region of our own Galaxy. But it was not until 1924 that the first galaxy outside our own was identified. Discovery has been added to discovery since that time, giving humankind a totally new picture of the universe. Galaxies ranging greatly in size, grouped together in clusters and even superclusters, pervade all visible space. Our own place in the universe doesn't seem particularly outstanding. The Milky Way galaxy, though not a dwarf, is not of unusual size. The Local Group, a cluster of galaxies which includes the Milky Way, is not as impressive as the great clusters that astronomers study.

In this lesson you'll learn how galaxies are studied. You'll notice that the name Edwin Hubble is mentioned often during this lesson, for he not only identified the first galaxy, but went on to develop a comprehensive system of classification of galactic structure. Hubble also developed what is now known as the "Hubble law," a law that defines the relationship between the speed of recession (the speed at which a galaxy is moving away from ours) and its distance from us. You'll study some other methods that are used to determine distances to faraway galaxies (even the brightest of cepheid variables, you'll find, don't penetrate the distances we are considering now). The lesson also presents some theories about how galaxies are formed, how they occasionally interact, and how they age.

One of the more interesting experiments in recent years has been the development of computer simulations of galaxies in action. Millions of years of time are compressed into just a few minutes, so a computer can graphically depict how galactic nebulas may have condensed to form spiral or elliptical galaxies and show how such objects might interact during collisions. Some examples of computer simulations are discussed in the text and in the television program.

When you have completed this lesson, you probably will agree that in this aspect of astronomy, more than any other, the questions and uncertainties far outnumber the answers. What forces underlie the radio galaxies? How do we account for the various galaxy shapes? These and many other questions have only tentative answers, if any. The mystery, and the fascination, remain.

Learning Objectives

When you have completed all assignments in this lesson, you should be able to:

1. Describe the methods for determining galactic masses.

2. Name and describe the classification scheme used in distinguishing different types of galaxies.

3. Describe and explain the evolutionary significance of galactic classifications.

4. List and identify the types of extragalactic distance indicators.

5. Describe the computer simulations and actual observations showing that galaxies collide.

Assignments for This Lesson

Before Viewing the Program

Read the overview at the beginning of this lesson and look over the learning objectives that precede the assignments.

Read all of text Chapter 25 and the section "Formation and Evolution of Galaxies" in Chapter 26, pages 456-457.

View Program 27, "Galaxies"

As you view the program, look for:

- the description of the Hubble classification scheme of galaxies and the characteristics of each type of galaxy (spiral, elliptical, irregular).
- the explanation of the various techniques used to measure distances to galaxies (cepheids, globular clusters, bright stars, supernovas, etc.)
- the description of the content of our "Local Group" of galaxies.
- the description of the evolution of the different shapes of galaxies.
- the description of computer simulations to illustrate collisions between galaxies.

After Viewing the Program

Review the text assignments.

Review the terms that appear in italics or boldfaced type in the text. In particular, note the key terms and concepts listed in the study activities section.

Complete the study activities and the self-test in this study guide; then check your answers with the answer key at the end of the lesson.

Edwin Hubble observed many hundreds of galaxies and eventually came to classify them according to their shape and content. The revised *Hubble Sequence*, used to classify galaxies today, can be best illustrated in the form of a tuning fork diagram. The galaxies range from the most spherical elliptical galaxies (E) through the highly flattened spiral (S) and barred spiral galaxies (SB) to the irregular galaxies (Irr). The criteria used to classify galaxies include:

1. types of stars
2. relative abundance of interstellar gas and dust
3. the amount of flattening
4. tightness of the winding of the spiral arms
5. relative size of the central hub or bulge of the galaxy compared to the disk

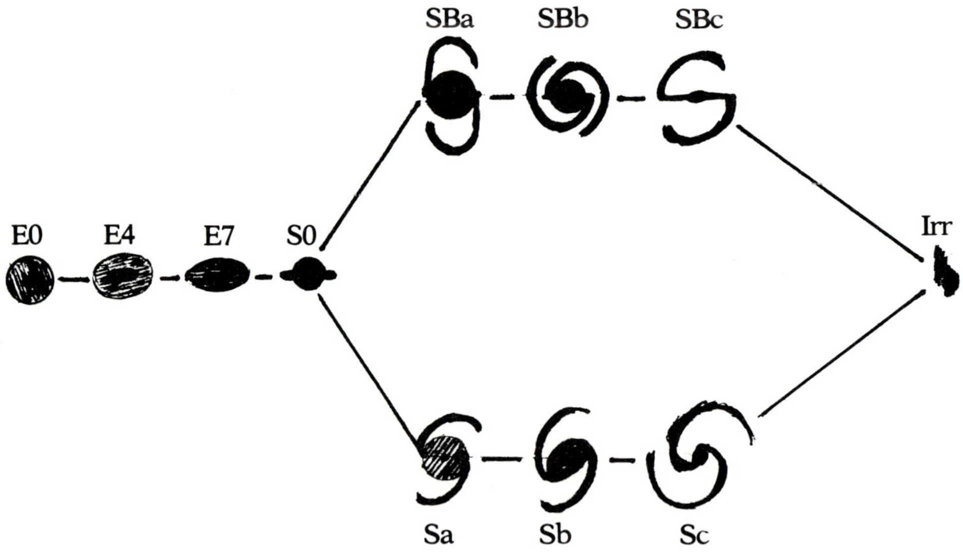

Figure 27.1. Hubble Sequence or "tuning fork."

Complete the following exercises: 2, 3, and 8 in text Chapter 25 on page 445.

Study Activities

Key Terms and Concepts

Hubble classification
spiral galaxy
elliptical galaxy
irregular galaxy
cannibalism (between galaxies)

Using What You've Learned

1. Using a source other than your text, make a chart of the shapes of the various types of galaxies according to the Hubble classification system. Find photographs of the following galaxies and classify them according to the Hubble scheme (most of these should be in your text):
   a. Andromeda M31
   b. M32 and NGC 205
   c. large and small Magellanic Clouds
   d. M51
   e. NGC 1300
   f. M81 (NGC 3031)
   g. M87 (NGC 4486)
   h. NGC 253

2. What is the major difference between spiral and elliptical galaxies?

Recommended Readings

For supplementary reading see the bibliography in the back of this study guide.

Self-Test

(Select the one best answer.)

Objective 1: Describe the methods for determining galactic masses.

1. In order to measure the mass of a spiral galaxy we must
   a. view it face on (perpendicular to our line of sight) and without any companions.
   b. view it for several days in order to measure its velocity through space.
   c. observe its gravitational influence on nearby galaxies.

2. After determining the total mass of the galaxy, and the average mass of a star from the mass luminosity relation and luminosity function, the ______________ in a galaxy can be determined.
   a. total amount of gas and dust
   b. total number of white dwarfs
   c. number of spiral arms
   d. total number of stars

Objective 2: Name and describe the classification scheme used in distinguishing different types of galaxies.

3. Dwarf ellipticals are the most abundant type of galaxy.
   a. true
   b. false

4. Irregular-shaped galaxies have the greatest percentage of their mass tied up in ______________ as compared with other types of galaxies.
   a. stars
   b. globular clusters
   c. gas and dust
   d. galactic clusters

5. Elliptical galaxies can be the most massive and contain the most stars.
   a. true
   b. false

6. Elliptical types of galaxies have the greatest percentage of ______________ compared to other types of galaxies.
   a. gas and dust
   b. old stars
   c. galactic clusters
   d. young stars

Objective 3: Describe and explain the evolutionary significance of galactic classifications.

7. The evolutionary sequence of galaxies, beginning with the youngest, is
   a. irregular, spiral, elliptical.
   b. irregular, elliptical, barred spiral, spiral.
   c. elliptical, spiral, irregular.
   d. There is no evolutionary sequence among galaxies; all were formed at about the same time.

Objective 4: List and identify the types of extragalactic distance indicators.

8. The parallax of a galaxy can be used to determine its distance.
   a. true
   b. false

9. It was always known that the "spiral nebulas" were extragalactic objects.
   a. true
   b. false

Objective 5: Describe the computer simulations and actual observations showing that galaxies collide.

10. Collisions between galaxies
    a. never occur
    b. occur and have no effect on the two galaxies.
    c. may be one of the explanations for the many peculiar galaxies that have been observed.

11. Because of the large numbers of stars in a galaxy, it is impossible for computers to simulate what would happen when two galaxies collide.
    a. true
    b. false

12. Evidence for the existence of colliding galaxies may have been observed in objects identified as
    a. quasars.
    b. black holes.
    c ring galaxies.
    d. Seyfert galaxies.

13. The distances to the nearest galaxies were first determined by using what type of variable star?
    a. Mira
    b. cepheid
    c. RR Lyrae
    d. pulsating

Answer Key

1. c	8. b
2. d	9. b
3. a	10. c
4. c	11. b
5. a	12. c
6. b	13. b
7. d	

Quasars 28

Overview

Do you find some of the concepts and distances in astronomy challenge your imagination? In this lesson you'll be considering a type of object that's still a challenge to everyone's imagination.

Recall, if you will, the size of the solar system. You may remember that the sun's light takes about five hours to arrive at Pluto's orbit. Now imagine an object just about seventy times (about one "light-month") larger in diameter than the total diameter of the solar system. That's big, but not quite one-fiftieth of the distance from the sun to nearby Alpha Centauri. By now you're aware that even an object of such size would be a tiny speck in the Milky Way. However, imagine this object radiating as much energy as 100 or more Milky Way galaxies — a hundred times more luminous than our Galaxy or the one in Andromeda. That seems to be the best picture available right now of "quasi-stellar radio objects," or quasars.

Unlike black holes, quasars were not envisioned in theoretical concepts before they were discovered. And there has been no simple or satisfactory answer yet presented which accounts for all the behaviors and properties of these strange objects.

In fact, there is a fair amount of controversy over quasars. As you'll learn in the course of this lesson, their luminosity is computed from estimates of their distance. The distance of the quasars, in turn, is derived from Hubble's law and the redshift in their spectra. If quasars were actually closer than their redshifts indicate, their absolute magnitudes wouldn't have to be as great to account for their apparent luminosities. Many astronomers have attempted to account for the redshift in some way, and some of their theories are presented in this lesson. But no proposed theory of "nearby quasars" has really been satisfactory either.

This lesson describes how quasars were discovered and how the properties of quasars have been determined. Studying them is a challenge, for they are among the most distant objects observed in the universe (the brightest of them is much too faint to be seen with the naked eye). You'll also learn about some galaxies which may be related to quasar phenomena: radio galaxies, Seyfert galaxies, and other galaxies with very active nuclei. This lesson also explores more of the theory of black holes, because some behavior of supermassive black holes could account for quasar energy production.

The mystery of the quasars and the controversy surrounding them are excellent examples of the scientific method in action. Obviously, more information needs to be gathered, but that very search for the new data is a challenge when the phenomena are so close to the edge of the detectable universe. It's likely that someone will devise ways to collect new data, or analyze current data in new ways. It's even possible that some new and unexpected solutions to the questions about quasars will be found. One thing certain, that new questions with difficult-to-find answers will arise as long as humankind looks toward the sky.

Learning Objectives

When you have completed all assignments in this lesson, you should be able to:

1. List and describe the types of characteristics of radio, Seyfert, and peculiar galaxies.

2. Describe the unique observational characteristics of quasars.

3. Explain the competing theories of the nature of quasars.

4. Describe the characteristics of rotating black holes.

5. List the evidence suggesting the existence of supermassive black holes in the cores of galaxies.

6. Describe the characteristics of a gravitational lens.

Assignments for This Lesson

Before Viewing the Program

Read the overview at the beginning of this lesson and look over the learning objectives that precede the assignments.

Read all of text Chapter 26.

View Program 28, "Quasars"

As you view the program, look for:

- a description of the nature and physical characteristics of quasars.
- an explanation of the apparent superluminal (faster than light) velocities observed in quasars.
- a description of the nature and physical characteristics of radio, Seyfert, and peculiar galaxies (M82).
- an explanation of the evidence for the existence of supermassive black holes in the cores of some galaxies.
- a description of the characteristics of the giant elliptical galaxy (M87).
- a description of the nature and discovery of gravitational lenses.

After Viewing the Program

Review the text assignments.

Review the terms that appear in italics or boldfaced type in the text. In particular, note the key terms and concepts listed in the study activities section.

Complete the study activities and the self-test in this study guide; then check your answers with the answer key at the end of the lesson.

As we studied the planets, stars, and galaxies, we discovered that they all shared a common property — rotation. It was therefore proper for the theoretical astronomer to consider the possibility that black holes also rotate. It wasn't until 1963 that Roy P. Kerr discovered the solution to the field equations of general relativity that included the property of rotation. This new solution, called the Kerr solution or the Kerr black hole, revealed many new and exciting characteristics of the black hole. The details of these new aspects of the rotating black hole are too complex to be completely described here. For further information, please read one of the references on black holes mentioned in Lesson Twenty-six of this study guide.

There is one feature of the Kerr black hole that needs to be mentioned now, because it may explain a bizarre phenomenon associated with several galaxies and quasars. Look at the photographs of the galaxy M87 (Figure 26.4, text page 451) and the quasar 3C 273 (Figure 26.3, text page 449). These two photographs show "cosmic jets" emanating from the centers of the two objects. The following is a possible explanation for the origin of the jets. (Your text describes another.)

According to the Kerr solution, there is a region around the outside of a rotating black hole in which one cannot remain at rest. In other words, spacetime is being whirled around the black hole and no matter what is attempted (even traveling at the speed of light), one cannot avoid being spun around the black hole. This region is called the ergosphere.

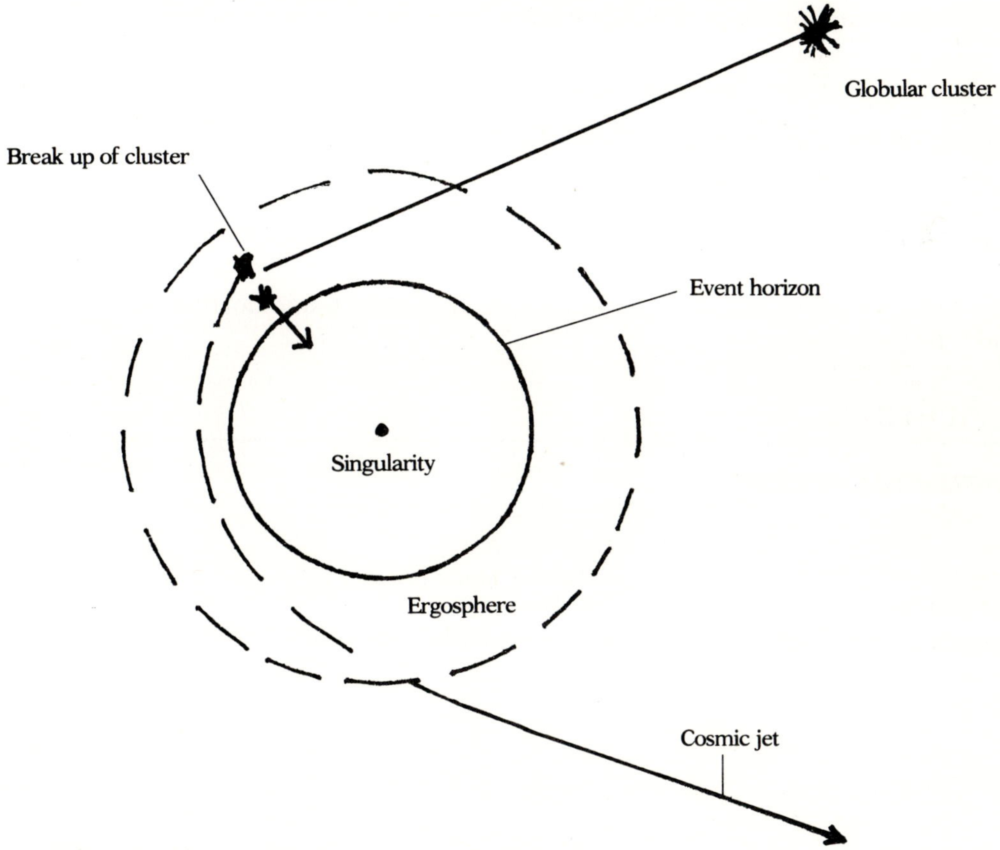

Figure 28.1. Rotating black hole.

Since the ergosphere is above the event horizon, any object caught within this whirlpool of spacetime is not doomed to fall into the black hole itself. Objects can be ejected from the ergosphere and ejected with much greater amounts of energy than they had when they entered. This fact could explain the cosmic jets.

Globular clusters have been observed around the galaxy M87. It is possible that a globular cluster was caught in the ergosphere and torn apart by gravitational forces with a portion of the cluster falling into the hole itself and the rest being ejected as the cosmic jet. Observations also reveal an extremely large dark mass, on the order of several billion suns, at the center of M87. Some astronomers have proposed that this dark mass is a supermassive black hole. If these conclusions are proved correct, then the general theory of relativity has provided us with more than just some mathematical exercises.

Complete the following exercises:

1. Describe the characteristics of rotating black holes and compare them to a Schwarzschild black hole.

2. Describe the observed characteristics of quasars and the problems scientists are having in trying to explain their nature.

3. Give an explanation for the superluminal velocities observed in some quasars.

4. Explain the possible connection between quasars, Seyfert, and normal spiral type galaxies.

Study Activities

Key Terms and Concepts

cosmic jets
ergosphere
gravitational lens
peculiar galaxy (M82, M87)
quasar
radio galaxy
rotating black holes
Seyfert galaxy
superluminal velocity

Using What You've Learned

1. Seyfert galaxies, N galaxies, and quasars are all thought to have what in common?

2. Explain the Penrose mechanism associated with rotating black holes. You may want to read more about this subject in William J. Kaufmann III, *The Cosmic Frontiers of General Relativity* (Boston: Little, Brown, 1977).

3. Use additional sources to describe the observations and evidence that led some astronomers to conclude that there is a supermassive black hole in the galaxy identified as M87.

Recommended Readings

For supplementary reading see the bibliography in the back of this study guide.

Self-Test

(Select the one best answer.)

Objective 1: List and describe the types and characteristics of radio, Seyfert, and peculiar galaxies.

1. The radio energy emitted by a radio galaxy has been seen to originate from regions, called lobes, on either side of the galaxy and not necessarily from the visible object itself.
   a. true
   b. false

2. Seyfert type galaxies emit the greatest portion of their total energy output in the form of____________ radiation.
   a. X-ray
   b. visible
   c. radio
   d. infrared

Objective 2: Describe the unique observational characteristics of quasars.

3. Which of the following observations are associated with quasars?
   a. very large redshifts
   b. starlike appearance
   c. energy emitted from a region about a light-year in size
   d. all of the above

4. The number of quasars ____________ with increasing distance from Earth.
   a. increases
   b. decreases
   c. remains the same

5. If quasars are the exceedingly active nuclei of galaxies, and are at cosmological distances, then we are looking at galaxies that are
   a. very old and in the process of exploding.
   b. very much like our own Milky Way but with more active nuclei.
   c. very young and in the process of forming.
   d. colliding.

Objectives 1 and 2

6. It is impossible to observe quasars and Seyfert galaxies with the naked eye.
   a. true
   b. false

Objective 3: Explain the competing theories of the nature of quasars.

7. If Seyferts, quasars, and normal spiral galaxies are connected in an evolutionary sequence, then the sequence would go, from youngest to oldest,
   a. Seyferts, quasars, spiral galaxies.
   b. quasars, Seyferts, spiral galaxies.
   c. spiral galaxies, Seyferts, quasars.
   d. spiral galaxies, quasars, Seyferts.

Objective 4: Describe the characteristics of rotating black holes.

8. The ergosphere is a region around a rotating black hole in which one
   a. can remain at rest.
   b. can never remain at rest.
   c. can remain at rest only if traveling at the speed of light.
   d. must eventually cross the event horizon and collide with the singularity.

Objective 5: List the evidence suggesting the existence of supermassive black holes in the cores of galaxies.

9. Which of the following are characteristics of the galaxy identified as M87?
   a. giant elliptical galaxy
   b. jet of matter ejected from the nucleus of the galaxy
   c. possible supermassive black hole at the core
   d. all of the above

10. The mass of a black hole in the cores of some galaxies such as M87 has been estimated to be
    a. 10 solar masses.
    b. 100 solar masses.
    c. 1,000 solar masses.
    d. millions and possibly billions of solar masses.

Objective 6: Describe the characteristics of a gravitational lens.

11. At the moment, there is absolutely no evidence for the existence of a gravitational lens.
    a. true
    b. false

Answer Key

1. a	7. b
2. d	8. b
3. d	9. d
4. a	10. d
5. c	11. b
6. a	

The Expanding Universe 29

Overview

Remember the "stellar models" that were discussed in the lessons about the properties of stars? Astronomers construct such models to account for all observed characteristics while, at the same time, providing for all the known natural laws. Some of the known natural laws include the conservation of matter and energy and the conversion of matter into energy. Constructing a workable model of the universe is much the same. It has to fit what has been observed, it must follow natural laws and, to be scientific, it must lend itself in some way to further testing by observation — it must be able to predict future events.

This lesson (and the next) will acquaint you with the generally accepted hypotheses and theories of the large-scale structure of the universe and the changes continually taking place within it. You'll find there's quite a bit of uncertainty in this portrait of the universe despite the many discoveries made in this century. In broad outline, the currently accepted picture is that of an expanding universe, uniformly filled with matter, which began with a single fantastic explosion perhaps thirteen to twenty billion years ago. Even this picture is liable to change, though, because it is based on some unproven assumptions. One of the assumptions concerns the Hubble law, which you read about in Lesson Twenty-seven. As you probably remember, according to Hubble the distance of an object is in proportion to the velocity at which it moves away from an observer. Although this law has been verified, the exact value of the Hubble constant (the mathematical proportion itself) is not known with certainty, therefore making an estimate such as the age of the universe equally uncertain.

As you know, the fact of the expanding universe is generally accepted. There is another principle that seems almost as certain, the cosmological principle, which states that space is basically the same everywhere. In other words, the conclusions we draw about the shape and nature of the detectable universe are consistent with the description of the entire universe. The extent to which this principle has been confirmed is one of the more interesting topics of study in this lesson.

You've already learned that the galaxies often appear grouped in clusters of widely varying shapes and sizes. A deeper inquiry into the distribution of galaxies reveals that clusters themselves are grouped into still larger clusters, so that the observed uniformity of the distribution of matter as expressed by the cosmological principle would be true on a large scale, but not necessarily in a "local" region. In your reading, you'll find some conjecture and suggestive evidence indicating that the superclusters themselves are arranged into comparatively narrow lines or "filaments." These filaments may give the universe an appearance something like a three-dimensional spider web.

This lesson introduces you to the contrasting concepts of a static or steady state (continuous creation) universe and a nonstatic (big bang) universe. You'll learn, too, why it is believed that we have penetrated to the limits of the observable universe, even though the universe itself is limitless.

Learning Objectives

When you have completed all assignments in this lesson, you should be able to:

1. Describe the "Local Group" of galaxies.
2. List and describe the properties and distribution of clusters of galaxies and superclusters.
3. Describe the observations of Vesto Slipher and Edwin Hubble.

4. Understand and explain the significance of the Hubble law of recession and the Hubble constant.

5. Explain how the Hubble law is used to determine the distance to a galaxy and be able to use that law in a simple problem.

6. Define and explain the significance of the cosmological principle and how it differs from the perfect cosmological principle.

7. Explain the significance of the expansion of the universe.

8. Describe the big bang and steady state theories of the origin of the universe.

9. List and describe the various observational tests and their results used in distinguishing between the big bang and steady state cosmologies.

Assignments for This Lesson

Before Viewing the Program

Read the overview at the beginning of this lesson and look over the learning objectives that precede the assignments.

Reread carefully the sections on the Hubble law in text Chapter 25, pages 433-436.

Read carefully the section "The Distribution of Galaxies in Space," pages 440-444 of Chapter 25.

Skim pages 464-476 of text Chapter 27.

View Program 29, "The Expanding Universe"

As you view the program, look for:

- a description of the distribution of galaxies, clusters of galaxies, and superclusters, and the content of the Local Group of galaxies.

- a description of the observations of the redshift of galaxies made by Edwin Hubble; and an explanation of the Hubble law and the meaning of the Hubble constant.

- an explanation of the significance of the cosmological principle and the perfect cosmological principle.
- a presentation of the big bang and steady state theories of the origin of the universe.
- the observations leading to the discovery of cosmic background radiation and an explanation of its significance.
- a description of Albert Einstein's initial understanding of the nature of the universe.

After Viewing the Program

Review the text assignments.

Review the terms that appear in italics or boldfaced type in the text. In particular, note the key terms and concepts listed in the study activities section.

Complete the study activities and the self-test in this study guide; then check your answers with the answer key at the end of the lesson.

Complete the following exercises: 2, 4, and 6 in text Chapter 27, page 479.

Study Activities

Key Terms and Concepts

Hubble law
Hubble constant
cosmic background radiation (CBR)
cosmological principle
clusters of galaxies
superclusters (distribution of observed matter)
Local Group
cosmology
big bang theory
steady state theory
perfect cosmological principle
cosmological constant

Using What You've Learned

1. Assume that the Hubble law is exact and that a galaxy that is 100 million parsecs away has a radial velocity of recession of 5,000 km/sec. A galaxy that has a radial velocity of 25,000 km/sec should be how many parsecs away?

2. What is the approximate age of the universe as we estimate it today?

3. Outline the differences and similarities between the steady state and big bang theories.

4. What is the major argument for rejecting the steady state theory?

Recommended Readings

For supplementary reading see the bibliography in the back of this study guide.

Self-Test

(Select the one best answer.)

Objective 1: Describe the "Local Group" of galaxies.

1. The Local Group contains
   a. spiral galaxies.
   b. elliptical galaxies.
   c. irregular galaxies.
   d. all of the above.

Objective 2: List and describe the properties and distribution of clusters of galaxies and superclusters.

2. Galaxies are distributed uniformly through space.
   a. true
   b. false

3. Recent observations of galactic clustering reveal that there exist
   a. only elliptical galaxies in clusters.
   b. only spiral galaxies in superclusters.
   c. large regions in space that seem to be nearly void of galaxies.
   d. no such phenomena as superclusters.

Objectives 3 and 4: Describe the observations of Vesto Slipher and Edwin Hubble. Understand and explain the significance of the Hubble law of recession and the Hubble constant.

4. The suggestion that the universe is expanding was first proposed as a result of the discovery of the
   a. Hubble law.
   b. Doppler effect.
   c. Stefan-Boltzmann law.
   d. general theory of relativity.

Objectives 4 and 5: Understand and explain the significance of the Hubble law of recession and the Hubble constant. Explain how the Hubble law is used to determine the distance to a galaxy and be able to use that law in a simple problem.

5. Galaxy A has twice the recessional velocity of galaxy B. According to the Hubble law, galaxy A is
   a. at the same distance as galaxy B.
   b. half the distance of galaxy B.
   c. twice the distance of galaxy B.
   d. four times the distance of galaxy B.

6. A galaxy has a recessional velocity of 10,000 km/sec. If the Hubble constant is 50 km/sec per megaparsec, what is the distance to the galaxy?
   a. a thousand pc
   b. 20 million pc
   c. 200 million pc
   d. 2 billion pc

Objective 6: Define and explain the significance of the cosmological principle and how it differs from the perfect cosmological principle.

7. The idea that the universe looks the same from all places is an expression of the
   a. cosmological principle.
   b. general theory of relativity.
   c. perfect cosmological principle.
   d. special theory of relativity.

Objective 7: Explain the significance of the expansion of the universe.

8. The big bang theory seems to account for the observed expansion of the universe. If this theory is correct, we exist in an evolving and changing universe.
   a. true
   b. false

Objective 8: Describe the big bang and steady state theories of the origin of the universe.

9. The big bang theory does not allow the contents of the universe and the relationships between the contents to evolve (change) with time.
   a. true
   b. false

10. The steady state theory assumes the validity of the ____________ and the big bang assumes the validity of the ____________.
    a. cosmological principle, perfect cosmological principle
    b. perfect cosmological principle, cosmological principle
    c. They both assume the validity of the cosmological principle.
    d. They both assume the validity of the perfect cosmological principle.

11. Both the big bang and steady state theories of the universe allow for the expansion of the universe.
    a. true
    b. false

Objective 9: List and describe the various observational tests and their results used in distinguishing between the big bang and steady state cosmologies.

12. Today the cosmic background radiation has a temperature of about how many degrees Kelvin?
    a. 3
    b. 300
    c. a million
    d. a trillion

13. Which of the following discoveries is totally consistent with the big bang cosmology and inconsistent with the steady state cosmology?
    a. expansion of the universe
    b. location in space where the big bang occurred
    c. cosmic background radiation
    d. blueshift in the velocities of distant galaxies

Answer Key

1. d
2. b
3. c
4. a
5. c
6. c
7. a
8. a
9. b
10. b
11. a
12. a
13. c

The Big Bang 30

Overview

In this last lesson you'll explore the event that may acount for all the matter and energy in the universe. In the middle of the 1960s, all speculation concerning ultimate beginnings and endings of the universe could be called, at best, hypotheses. Now, certain observations have taken one model out of the realm of conjecture and into the realm of theory. One tremendous explosion of an unimaginably dense mass seems to have left, as if in its wake, a few detectable traces that inspire more and more scientific inquiry. If the theory of the big bang continues to gain verification from further observations, it will become one of the great achievements of modern astronomy.

Somewhat like black holes, the big bang model has achieved a certain measure of popularity. Many books and articles covering just the first few minutes of the formation of the universe have been written for the lay reader. As you'll discover in this lesson, some scientists devote their attention and inquiries to the first seconds and fractions of seconds of the big bang. As you might expect, this theory is not neatly bound in a simple package, but contains some concepts that challenge the imagination. You'll probably find it almost impossible to visualize some of these concepts except by analogy, yet it is well worth the effort.

Can you imagine what may someday result from a scientific understanding of the origin of the universe? To many people, increasing basic knowledge is valuable enough; knowledge, like virtue, is its own reward. But pure research, such as America's experience with its space program, can have "spin-offs" that affect the lives of everyone. For example (while this is only speculation), it is possible that someday the knowledge of the first moments of creation would lead to a new understanding of the nature of matter, leading in turn to new ways of manipulating matter and energy.

In this lesson you'll learn that scientists explore still other points of view in attempts to answer some basic questions. You'll find out how the big bang theory fits in with some important observations. There are still other scientific searches underway for the primordial black holes which could have been spawned in the universe's first moments. There are many unanswered questions: Will the universe continue forever to expand, or will gravity eventually drag everything back to the "big crush" of a collapsed universe? There is much uncertainty revolving around what might be called "the mystery of the missing mass." Both the television program and the text give several examples of the importance of the "early universe," those first few seconds after expansion began. If the big bang theory is correct, the events in those first moments set the stage for all that has transpired in the ten to twenty million years since. Finally, you will be introduced to some theories not yet verified, which could resolve some of the problems posed by the big bang theory.

It is a long journey, indeed, from the good Earth and its moon, where **Project Universe** began, to the edge of the observable universe and the theoretical-but-possible beginnings of space and time. One of the many goals of this course is to introduce you to things of wonder and of beauty throughout the known universe. We hope you have found some of them. The potential for new knowledge is everywhere. To understand the universe we must begin with the ground beneath us. To understand the basic forces controlling matter, we may have to search the limits of space and time.

Learning Objectives

When you have completed all assignments in this lesson, you should be able to:

1. Describe what the big bang cosmology predicts regarding the evolution and contents of the universe.

2. Describe the suspected nature of primordial (mini) black holes and their implications for cosmology.

3. Describe the geometric and physical differences between open and closed universes.

4. Describe how the observations of the deuterium abundance, total mass of the universe, and temperature of the big bang are being used to distinguish between an open and a closed universe.

5. List and describe the other plausible cosmologies involving alternate explanations for the redshift, the relationships of various physical constants, and the origin of the universe.

6. Describe the nature and characteristics of the "very early" universe.

7. Describe and explain the significance of the grand unified theories (GUTS), supersymmetry model, and the inflationary universe model.

Assignments for This Lesson

Before Viewing the Program

Read the overview at the beginning of this lesson and look over the learning objectives that precede the assignments.

Read all of text Chapter 27. Also read the section "Other Ideas about Black Holes," pages 405-406 of Chapter 23.

View Program 30, "The Big Bang"

As you view the program, look for:

- a description of the nature and characteristics of primordial black holes.
- an explanation of the nature of the observations used to distinguish between an open and a closed universe.
- the geometry, predicted by general relativity, of an open and of a closed universe.
- an explanation of what is meant by the "missing mass" and the importance of the discovery of a possible neutrino mass.

- a description of the nature of the "very early" universe.
- the significance of the grand unified theories (GUTS), supersymmetry model, and the inflationary universe model.

After Viewing the Program

Review the text assignments.

Review the terms that appear in italics or boldfaced type in the text. In particular, note the key terms and concepts listed in the study activities section.

Complete the study activities and the self-test in this study guide; then check your answers with the answer key at the end of the lesson.

Complete the following exercises: 1, 2, and 6 in text Chapter 27, page 479.

Study Activities

Key Terms and Concepts

anisotropy of the CBR
primordial black holes
open universe
closed universe
deuterium
neutrino mass
negatively curved universe
positively curved universe
missing mass
flat universe
grand unified theories (GUTS)
inflationary universe

Using What You've Learned

1. Under the open universe theory, how will the overall large-scale density of the universe change over time?
2. Suppose the closed universe big bang theory is correct. You are an astronomer on a planet observing the spectra of galaxies only 2 billion years before the eventual catastrophe. What should you expect to notice about the spectra?
3. Outline the differences between primordial black holes and stellar mass black holes.

nded Readings

entary reading see the bibliography in the back of this study

---

e one best answer.)

1: Describe what the big bang cosmology predicts regarding the evolution and contents of the universe.

he center of the universe is located in the direction of what onstellation?
   a. Sagittarius
   b. Cygnus
   c. Andromeda
   d. There is no center of the universe.

bjective 2: Describe the suspected nature of primordial (mini) black holes and their implications for cosmology.

2. Stephen Hawking predicts that primordial black holes would be observed as a source of
   a. radio waves.
   b. visible light.
   c. gamma rays.
   d. infrared radiation.

3. Mini black holes are predicted to
   a. evaporate and explode.
   b. grow in size and become stellar mass black holes.
   c. reach a minimum size and then remain at this size for eternity.

Objective 3: Describe the geometric and physical differences between open and closed universes.

4. The geometry of an open universe would be
   a. positively curved.
   b. negatively curved.
   c. spherical.
   d. flat.

Objective 4: Describe how the observations of the deuterium abundance, total mass of the universe, and temperature of the big bang are being used to distinguish between an open and a closed universe.

5. Which of the following would be an observation of a closed universe.?
   a. overabundance of deuterium
   b. scarcity of deuterium
   c. determination of the direction of the center of the universe
   d. indefinite expansion

6. The current observations of the total mass of the universe, deuterium abundance, and deceleration parameter indicate that the universe is
   a. closed and spherical in shape.
   b. closed and negatively curved.
   c. open and negatively curved.
   d. open and spherical in shape.

Objective 5: List and describe the other plausible cosmologies involving explanations for the redshift, the relationships of various physical constants, and the origin of the universe.

7. A change in the masses of subatomic particles, used to explain the redshift-distance relationship of the Hubble law, is a component of
   a. Lemaitre's big bang.
   b. Alpher, Bethe, and Gamow's theory of 1948.
   c Hoyle's revised steady state theory.
   d. antimatter cosmologies.

Objective 6: Describe the nature and characteristics of the "very early" universe.

8. It is believed that during a period of time very early in the history of the universe the four forces of nature were
   a. unified.
   b. distinct, but different from what they are today.
   c. distinct, but the same as they are today.
   d. unified, but separated within the last billion years.

Objective 7: Describe and explain the significance of the grand unified theories (GUTS), supersymmetry model, and the inflationary universe model.

9. A model of the early universe includes a period of rapidly accelerating expansion. This model is called the
   a. big bang.
   b. steady state.
   c. antimatter cosmology.
   d. inflationary universe.

10. The current attempts to unify the forces of nature are expressed in what are known as grand unified theories and have been completely successful.
    a. true
    b. false

Answer Key

1. d	6. c
2. c	7. c
3. a	8. a
4. b	9. d
5. a	10. b

Appendix 1: Answer Key

Following are the answers to questions from the section "Using What You've Learned."

Lesson 1

1. 1.a, and 1.b. In these activities, you will notice change. Over the course of the semester you will notice that the position of the rising and setting sun will drift north and south along the horizon. At the sun's northernmost and southernmost positions are what are called the solstices (the longest and shortest days, respectively, as seen in the northern hemisphere). When the sun rises due east and sets due west, the equinoxes occur, usually on the first days of autumn and spring.

   The constellations will also change; the stars rise four minutes earlier every day. Therefore at 9:00 P.M. each night the stars will be four minutes ahead of where they were the previous night. By the end of the semester you will notice a completely different set of constellations overhead.

Lesson 2

1.

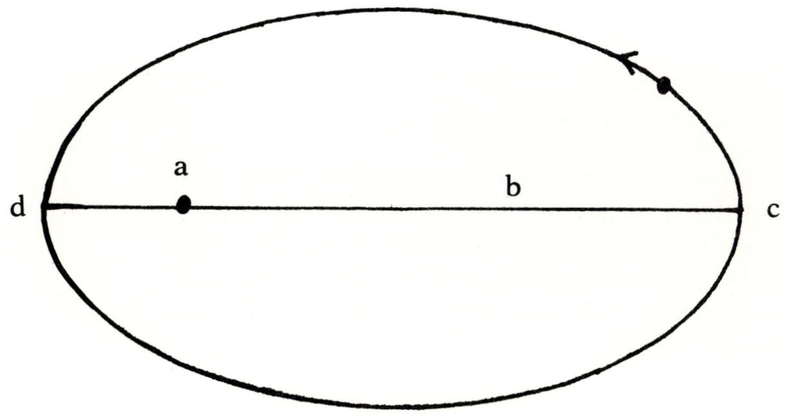

Figure 2.2. Planetary orbit.

2. The astronaut moves in a direction opposite to the direction in which he or she threw the wrench. If, as we expect, the astronaut is more massive than the wrench, his or her speed is less than that of the wrench, since speed is inversely proportional to mass.

3. a. Nowhere.
   b. At the equator.
   c. At the Artic Circle.
   d. Everywhere.
   e. At the Tropic of Cancer.
   f. Everywhere in the northern hemisphere.
   g. At the South Pole.
   h. At the equator.

Lesson 3

3. Some sources that contain information on eclipses are the *American Ephemeris, Astronomy,* and *Sky and Telescope.*

5. The angular diameters of the sun and moon are too small for the total phase of a solar eclipse to last an hour. Also, the moon would have to be much larger or nearer for an eclipse to last an hour. The maximum time for a solar eclipse is about seven minutes.

Lesson 4

1. 1) b
   2) $c = \lambda f$
   3×10^{10} cm/sec $= \lambda$ (5.00×10^{14} cycles/sec)
   6×10^{-5} cm $= \lambda$ or
   6,000 Å $= \lambda$ where Å $= 10^{-8}$ cm diameter of hydrogen atom
   3) b
   4) a

2. The brightness of a 100-watt lamp at 50 meters equals the brightness of a second lamp at 100 meters. The second lamp is twice as far away ($100/50 = 2$) as the 100-watt lamp. Using the inverse-square law of light, the second lamp must be 2^2 times brighter than the 100-watt lamp in order for it to appear the same brightness. If $2^2 = 4$, the second lamp is 4×100 watts or 400 watts.

Lesson 5

2. a. Level 1
   b. More energy is required to raise the electron to level 3.
   c. The change from (1) to (2).
   d. The change from (1) to (3).

3. a. refractor
   b. reflector
   c. reflector
   d. reflector
   e. refractor

Lesson 6

1. See Figure 7.2 in your text on page 128.

2. a. Small particles in the atmosphere scatter sunlight. Since blue wavelengths of light are scattered the best, the sky appears to be blue.

   b. This effect is an illusion. The sun is the same size at both times. When it is on or near the horizon, you have something with which to compare it, such as trees, mountains, and the like.

   c. Halos form around the sun and the moon because ice crystals in the atmosphere cause the sunlight and moonlight to be redirected, refracted, toward Earth. The same thing happens when light is passed through a chandelier.

4. 300,000 km. To determine the answer, use the formula $D = ct$, where D = distance, C = the speed of light, and t = the time in seconds. Multiply the speed of light (300,000 km/sec) times the time (1 sec each way).
   $D = ct$
   $= (300{,}000 \text{ km/sec})(1 \text{ sec})$
   $= 300{,}000 \text{ km}$

Lesson 10

3. Based on observations made by the Pioneer and Voyager spacecraft, the Great Red Spot is believed to be a storm system that may have been around for a million years.

Lesson 12

1. Because its axis rotation is nearly lying in the orbital plane, each hemisphere of Uranus would more directly face the sun. This would probably cause each hemisphere of Uranus to have more extreme weather conditions than Earth.

3. Pluto's orbit is so elongated that a portion of it brings the planet within the orbit of Neptune. At the present time, Pluto is in that portion within Neptune's orbit and therefore is closer to the sun than Neptune. This condition should last until about the year 2000.

Lesson 14

2. To mention a few. it was reported that Europe suffered through a mini-ice age and what is now the Southwestern United States experienced a severe drought due to the Maunder minimum.

Lesson 16

	1	2	3	4	5
Distance (parsecs)	100	5,000	30	25	250
Parallax (sec)	.01	.0002	.033	.04	.004
Radial Vel (km/sec)	−60	+70	+3	−3	−10
Proper Motion (sec/yr)	1.20	1.0	1.5	2.5	1.0

1. 1, 2 respectively
2. 4
3. 2
4. 3
5. 4
6. 4, 2 respectively

Lesson 17

1. a, c (See also text Figures 16.8 and 16.9, page293.)

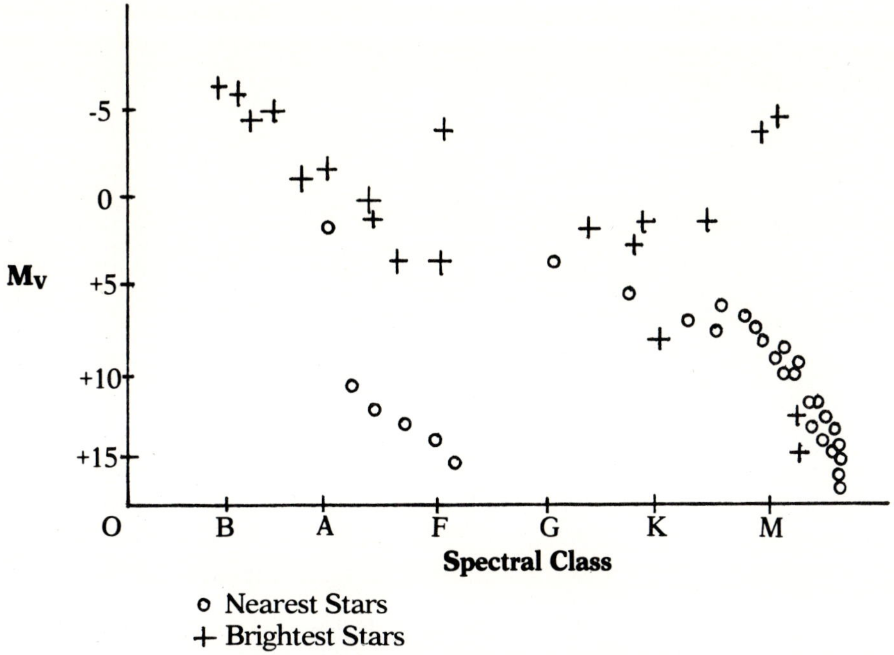

Figure 17.3. H-R diagram.

b. The nearest stars are generally small and faint and fall on the lower right end of the main sequence. The brighest stars fall at the upper left portion of the main sequence and also in the giant branch of the H-R diagram.

2. a-c
   b-e
   c-b
   d-d
   e-a
   f-b
   g-e
   h-e
   i-b
   j-d

Lesson 18

1. b
2. a
3. e
4. d
5. e
6. d
7. b
8. d
9. d
10. b
11. a

Lesson 19

1.

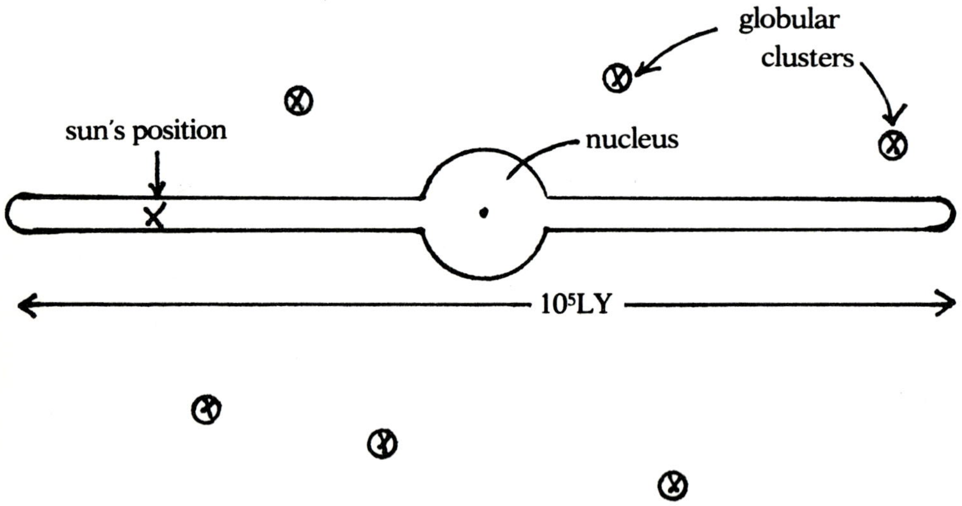

Figure 19.1. Side view of our Galaxv.

2. As hot HII gas pushes into the surrounding cold HI gas, the density builds up at the boundary; at such places stars might form.

3. Interstellar dust scatters blue light more than red light. So, red light passes more easily through interstellar dust than does blue light. Therefore, the stars partially obscured by interstellar dust appear redder than normal.

Lesson 20

1. e
2. d
3. a
4. c
5. c

6. Evidence in support of the existence of the magnetic field of the Milky Way can be found in synchrotron radiation, polarization of light, and charged particles moving in curved paths.

Lesson 21

1. Increasing the carbon dioxide content of the atmosphere would blanket the infrared radiation of Earth near the ground (greenhouse effect). The temperature would rise, polar ice caps would melt, and climate changes could be disastrous.

2. O, B, and A stars probably do not survive long enough for life and a civilization to arise. M stars are so faint that if the planet is to have reasonable temperatures it must be very close; however, then it can become locked in by gravitational tidal forces, resulting in one side that is always hot and the other, cold.

3. F, G, and K type stars have lifetimes long enough to allow life and civilizations to arise. Our sun is a G type star. The energy output of these stars is great enough so that planets may have reasonable temperatures at a great enough distance that gravitational tidal forces are not a problem.

4. ϵ Eridani
   ϵ Indi
   τ Ceti

Lesson 22

1. c
2. a
3. b
4. d

Lesson 23

1. T Tauri stars — very young stars, surrounded by gas and dust envelopes, associated with large quantities of infrared radiation, extremely strong stellar wind. These stars are losing significant amounts of matter.

   Bok globules — small, dark, roughly spherical blobs surrounded by light inside large glowing nebulas. Regions of dense stellar matter believed to be sites of future stellar formation.

   Herbig-Haro objects — strange, bright, and sometimes variable nebulas believed to be the sites of newly emerging stars in the very earliest stages of formation.

Lesson 24

1. a. These stars are located all along the main sequence.

   For the answers to b through g, see Figure 24.2.

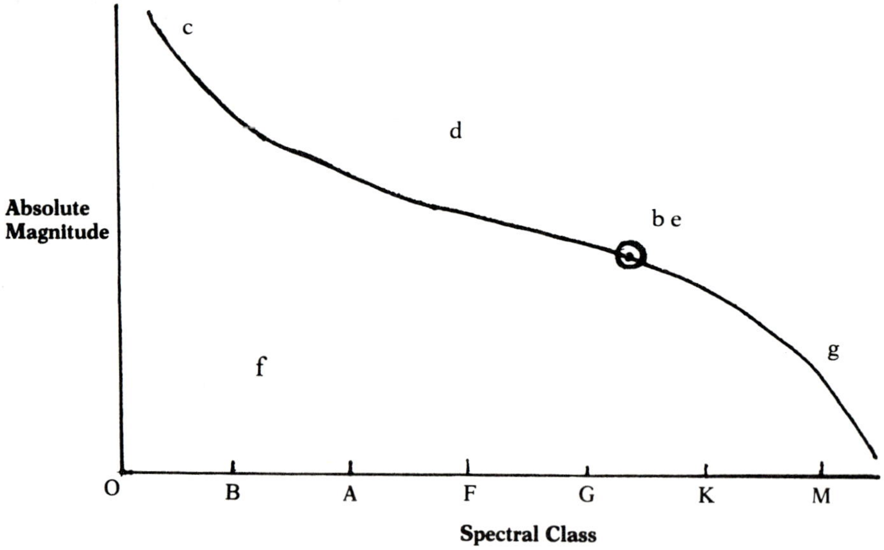

Figure 24.2. Star clusters.

2. D, C, A, B

Lesson 25

1. b, c, a, e, d

2. A nova event is the brightening of a star, believed to be a white dwarf, that is a member of a binary system. The brightening occurs as the result of the larger companion dumping material onto the white dwarf. This material is heated tremendously and burns away, resulting in a rapid brightening of the object.

   A supernova event is the brightening of a star caused by an explosion that destroys the star. A supernova event can happen only once, whereas a nova event may occur several times.

Lesson 27

1. See Figure 27.1 in this study guide.
   a. Sb
   b. E2 and E5 respectively
   c. Irr (Irregulars)
   d. Sc
   e. SBc
   f. Sb
   g. EO
   h. Sa/Sb

2. Spiral galaxies are highly flattened stellar systems (with a high angular momentum) shaped like large oval serving platters. They contain a variety of stars, ranging in age from newly forming and young to old and dying stars. A good percentage of the total mass of spiral galaxies is made up of interstellar gas and dust.

   Elliptical galaxies are spherical to oval stellar systems (with a low angular momentum) containing mostly older stars. Only a small percentage of the total mass of elliptical galaxies consists of interstellar gas and dust.

Lesson 28

1. They are all known for violent events in their centers. (This is what quasars are conjectured to be.)

2. The Penrose mechanism is a method by which energy may be extracted from a rotating black hole. An object that falls into the ergosphere around a black hole may be ejected with greater energy than it had when it entered the region. The energy gained by the object is lost by the black hole. Thus the hole rotates a little slower.

Lesson 29

1. The second galaxy is receding with a velocity five (5) times greater than the first galaxy. Therefore, according to the Hubble law, the second galaxy must be five times farther away.
   5 (100 million) = 500 million parsecs

2. The approximate age of the universe is 13 to 20 billion years.

Lesson 30

1. In the open universe, uniformly strong radio waves will be detected in all directions, corresponding to a black body of 3° K: this appears to be greatly redshifted radiation from the primeval fireball. The density of the universe should decrease, and slowly approach zero.

2. They should be shifted toward the violet (blue) end of the spectrum as the galaxies appear to move toward you. The apparent velocity of approach will be approximately proportional to their distance from you.

3. Primordial black holes were all formed at the time of the big bang; none are forming now. They are small in size (on the order of an electron) and have a mass similar to a mountain. They may evaporate or explode.

   Stellar mass black holes are large, many kilometers in diameter, and are forming today when stars collapse. They do not evaporate or explode.

Appendix 2: Significant Events in Astronomy

4977 B.C., April 27. Date of creation of the universe according to Johannes Kepler.

4004 B.C., October 22. Date of creation of the universe according to 1650 statement of Archbishop James Ussher.

780 B.C., June 4. First reliable recording of total solar eclipse by the Chinese.

648 B.C., April 6. Earliest Greek observation and recording of a total solar eclipse.

240 B.C., June 19. Circumference of Earth estimated by Eratosthenes.

1057, July 4. Chinese and Japanese astronomers observe supernova. Remnant later named the Crab nebula.

1604, October 8. Kepler's supernova observed.

1610, January 7. Discovery of first three satellites of Jupiter — Io, Europa, and Ganymede — by Galileo Galilei.

1610, January 13. Discovery of Callisto, fourth satellite of Jupiter, by Galileo Galilei.

1655, March 25. Discovery of Titan by Christian Huygens.

1687, July 5. Publication of *Principia* by Isaac Newton.

1705, April 16. Isaac Newton knighted by Queen Anne of England at Trinity College.

1780, May 2. Discovery of first binary star by William Herschel.

1781, March 13. Discovery of planet Uranus by William Herschel.

1801, January 1. Discovery of first asteroid, Ceres, by Giuseppe Piazzi.

1802, March 28. Discovery of second asteroid, Pallas, by Heinrich Olbers.

1803, April 26. Meteorites fall in France. Later shown to be from outer space by Jean Baptiste Biot.

1839, January 9. Measurement of the first stellar parallax by Thomas Henderson — Alpha Centauri.

1845, April 2. First successful photographs of the sun taken by H.L. Fizeau and J. Leon Foucault.

1846, September 23. Discovery of Neptune by J.G. Galle using the computations of Leverrier.

1846, October 10. Triton, largest moon of Neptune, discovered by William Lassell.

1864, August 29. Discovery of the chemical composition of the nebulas by William Huggens.

1868, August 18. Discovery of helium in solar spectrum by Norman Lockyer.

1877, August 11. Discovery of Deimos, moon of Mars, by Asaph Hall.

1877, August 17. Discovery of Phobos, moon of Mars, by Asaph Hall.

1908, June 30. Tunguska Event, Siberia. Believed to be explosion of a comet.

1910, April 21. Perihelion passage of Halley's comet.

1920, April 26. Curtis-Shapley debate over the nature of nebulas.

1923, October 5. Distance to Andromeda galaxy measured using a cepheid variable by Edwin Hubble.

1926, March 16. Launch of first liquid-fueled rocket by Robert Goddard.

1930, February 18. Discovery of Pluto by Clyde Tombaugh.

1930, May 12. First planetarium opens in the United States: Adler Planetarium, Chicago, Illinois.

1935, May 14. Opening of third planetarium in the United States: Griffith Observatory, Los Angeles, California.

1948, June 3. Dedication of 200″ telescope and Palomar Mountain Observatory.

1949, May 1. Discovery of Nereid, second moon of Neptune, by Gerard Kuiper.

1957, October 4. First artificial satellite launched by the Soviet Union — Sputnik 1.

1957, November 2. Soviet dog Laika — first Earth creature to orbit the planet.

1958, January 31. Launch of Explorer 1, first United States artificial satellite.

1958, July 29. National Aeronautics and Space Act signed by President Dwight D. Eisenhower.

1959, January 2. Launch of Mechta by Soviet Union. First lunar probe and object created by humans in orbit about the sun.

1960, April 6. Beginning of Project Ozma, first organized attempt to contact extraterrestrial intelligent life.

1960, August 12. Launch of first passive communications satellite by the United States — Echo I.

1961, April 12. First man to orbit Earth — Soviet cosmonaut Yuri Gagarin.

1961, May 5. First United States manned suborbital space flight completed by Alan Shepard.

1961, May 25. Goal set by President John F. Kennedy to land a man on the moon by the end of the decade.

1962, February 20. Completion of first United States manned orbital flight by John Glenn.

1962, March 7. Launch of the United States Orbiting Solar Observatory. First space observatory.

1962, July 10. Launch of first privately funded satellite — Telstar.

1962, August 27. Launch of Mariner II by the United States. First planetary probe to Venus.

1963, February 5. Discovery of the enormous redshifts in the spectra of the quasars by Maarten Schmidt.

1963, June 16. First woman in space — Soviet woman Valentina Tereshkova.

1964, July 31. U.S. Ranger crashes into moon after sending back over 4,000 photographs.

1964, November 28. Launch of Mariner 4, first spacecraft to send back photographs of Martian surface.

1965, February 17. Launch of U.S. Ranger 8 toward an impact in Mare Tranquilitatis. Took over 7,000 photographs of lunar surface before impact.

1965, March 18. First spacewalk by Soviet cosmonaut Aleksei A. Leonov from Voskhod II.

1965, March 21. Launch of U.S. Ranger 9 toward impact in Aphonsus Crater. Sent back over 5,000 photographs.

1965, March 23. Completion of first two-man space flight — three revolutions of Earth by Virgil Grissom and John Young in Gemini 3.

1965, April 6. Launch of first U.S. commercial satellite — "Early Bird."

1965, June 3. First U.S. astronaut space walk by Ed White in Gemini 4.

1965, December 15. First rendezvous in space performed by Gemini 6 and Gemini 7.

1966, February 3. First spacecraft to soft-land instrument package on the moon. Luna 9, Soviet Union.

1966, March 1. First artificial satellite to impact on another planet — Soviet Venus III.

1966, March 16. First docking of manned and unmanned vehicles by U.S. Gemini 8.

1966, March 31. Launch of Luna 10 by Soviet Union, first satellite to orbit the moon.

1966, June 2. First unmanned soft landing on the lunar surface — U.S. Surveyor I.

1966, August 10. Launch of U.S. Lunar orbiter 1 sends back first picture of Earth as seen from space.

1967, January 27. U.S. astronauts Virgil Grissom, Howard White, and Roger Chaffee die in fire during spacecraft launch rehearsal.

1967, April 19. U.S. Surveyor III soft-lands on lunar surface. Verifies surface is safe for landings.

1967, April 23. Launch of Soyuz I by the Soviet Union. First habitable space laboratory.

1967, October 18. First spacecraft to send back data from within atmosphere of Venus — Soviet Venera 4.

1968, February 24. Discovery of first pulsar announced by Cambridge astronomers.

1968, December 21-27. U.S. Apollo 8, first manned spacecraft in circumlunar orbit.

1969, January 16. First transfer of crew in space between Soviet ships Soyuz 4 and Soyuz 5.

1969, July 16. Launch of U.S. Apollo 11, first manned spacecraft to land on the moon.

1969, July 20. "The Eagle has Landed." First manned craft to land on the moon.

1969, July 21. Neil Armstrong and Edwin "Buzz" Aldrin make first walk on the moon.

1971, May 30. Mariner 9, first satellite to orbit Mars, launched.

1973, March 7. Discovery of Comet Kohoutek by Lubos Kohoutek, Hamburg Observatory.

1973, May 14. Launch of U.S. space station Skylab.

1974, March 29. Mariner 10 flyby of Mercury; photographs surface.

1974. November 16. Arecibo, 1,000 ft. (305 m) telescope sends message to globular cluster M13.

1975, October 22. Venera 9 lands on Venus. Sends back first photographs taken of the surface of another planet.

1976, July 20. Viking 1 lands on surface of Mars (Chryse Planitia). Later samples surface materials and sends back photographs.

1976, September 3. Viking 2 lands on surface of Mars (Utopia).

1977, March 10. Discovery of rings of Uranus.

1977, November 1. Discovery of asteroid Chiron by Charles Kowal.

1978, June 22. Discovery of Charon, moon of Pluto, by J.W. Christy.

1978, December 9. Four probes and mother ship reach surface of Venus — U.S. Pioneer Venus spacecraft.

1979, March 4. Discovery of ring around Jupiter by Voyager 1.

1979, March 9. Discovery of first active extraterrestrial volcano on Jupiter's moon Io.

1979, July 9. Voyager 2 spacecraft flyby of Jupiter. Confirms Voyager 1 findings.

1979, July 9. Disintegration of Skylab in atmosphere over Australia.

1979, September 1. Pioneer 11 first spacecraft to fly by Saturn. Detected two new rings labeled F and G.

1980, February 14. U.S. launches satellite — Solar Maximum Mission. First satellite to be equipped for retrieval by space shuttle.

1980, April 10. Quasar 3C 345 discovered by West German radio astronomers. Possibly the largest single object in the universe. Radio energy emitted from a region of space 78 million light-years across.

1980, November 12. Voyager 1 flyby of Saturn. Discovered: Titan's atmosphere to be mostly nitrogen; "ringlets" in the larger rings; "spokes" in the B ring.

1981, January 20. University of Wisconsin astronomers report discovery of star 3,500 times more massive than the sun, the most massive to date.

1981, April 12. Launch of first reusable spacecraft — space shuttle Columbia — for 36 Earth orbits. Crew of John W. Young and Robert L. Crippen.

1981, August 25. Flyby of Saturn by Voyager 2. Confirms initial discoveries of Voyager 1.

1981, October 2. Discovery by Kitt Peak National Observatory of huge void in space. Region could contain 2,000 galaxies but ovservations reveal area to be nearly empty.

1982, March 1. Soviet Venus probe Venera 13 lands on surface of Venus. Sends back eight photographs of planet's surface.

1982, March 10. Rare semi-alignment of planets; predicted to cause widespread seismic destruction in book titled the *Jupiter Effect.* No unusual events reported.

1982, November 11. Launch of space shuttle Columbia for first operational flight. Two satellites placed into orbit.

1983, January 25. Infrared Astronomy Satellite (IRAS) launched. Monitors infrared or heat radiation.

1983, April 4. First launch of second space shuttle, Challenger. Four-man crew: Paul Weitz, Karol Bobko, Story Musgrave, and Donald Peterson.

1983, June 13. Pioneer 10, launched in 1972, crosses the orbit of Neptune and becomes the first manned spacecraft to leave the solar system.

1983, June 18. Launch of space shuttle Challenger for second flight. Carries first U.S. female astronaut Sally Ride.

1983, August 9. Announcement by Jet Propulsion Laboratory of the discovery of a disk of large particles around the star Vega. Possibly the first direct observation of another solar system or merely a large belt of asteroid-sized objects.

1983, October 19. Subrahmanyam Chandrasekhar awarded Nobel Prize in physics for his work on the evolution of stars, especially white dwarfs (Chandrasekhar limit). The award was shared with William A. Fowler, who received his award for his work on a complete theory of the elements of our universe. (Synthesis of elements in the interior of stars and later ejected into space by the supernova event.)

1984, February 13. Announcement by JPL of the discovery of volcanic eruptions on the planet Venus by the Pioneer Venus probes in 1978. (It took 5 years to confirm data.)

1984, June 12. Discovery of ring of comets and belt of asteroids around the star Vega by IRAS (Infrared Astronomy Satellite).

1984, July 12. Possible ring around Neptune detected.

1984, December 15. Soviet Union launches Vega I space probe to rendezvous with Halley's comet in March 1986.

1985, June 7. Strong evidence for existence of black hole in the core of the Milky Way reported.

1985, July 2. Giotto spacecraft launched by the European Space Agency. Scheduled to rendezvous with Halley's comet in March 1986.

1985, September 11. U.S. spacecraft International Cometary Explorer crosses through the tail of Comet Giacobini-Zinner at a distance of 44 million miles from Earth.

1986, January 24. Voyager 2 closest encounter with the planet Uranus, at a distance of 50,600 miles.

1986, January 28. Explosion of the space shuttle Challenger. All crew members lost. Shuttle flights suspended.

1986, January 29. Eleventh ring of Uranus reportedly found.

1986, February 9. Comet Halley reaches perihelion.

1986, March 6. Vega I spacecraft penetrates the atmosphere of Halley's comet and reaches a distance of 5,6000 miles from the nucleus.

1986, March 9. Vega II penetrates the atmosphere of Halley's comet to within 5,125 miles of the nucleus.

1986, March 14. Giotto spacecraft comes within 335 miles of the nucleus of Halley's comet.

1986, April 28. Milky Way galaxy determined to be smaller than originally thought. New diameter determined to be 70,000 light-years.

1986, July 3. Most distant object in the observable universe discovered. A quasar at a distance of 12.4 billion light-years from Earth.

1987, February 24. Supernova discovered, closest to Earth since 1604, at a distance of 160,000 light-years (located in the Large Magellanic Cloud).

Appendix 3: Who's Who in Astronomy

Abell, George O. (1927-1983). Popularizer of astronomy, author of text for **Project Universe,** and discoverer of a number of superclusters.

Adams, John Couch (1819-1892). English astronomer and codiscoverer of Neptune.

Armstrong, Neil (b. 1930). First astronaut to walk on lunar surface.

Baade, Walter (1893-1960). Revised intergalactic distances upward by a factor of two (1952). Discoverer of asteroid Icarus, whose perihelion is within the orbit of Mercury.

Baily, Francis (1774-1844). English astronomer who described "Baily's beads" during total eclipse of the sun.

Bode, Johann (1747-1826). Popularized Titius rule of planetary distances, also known as "Bode's Rule."

Bradley, James (1693-1762). Discovered aberration of starlight.

Brahe, Tycho (1546-1601). Benefactor of Johannes Kepler. Gathered positional data of planets later used by Kepler to develop three laws of solar system. Built instruments for highly accurate astronomical measurements.

Bruno, Giordano (1548-1600). Italian philosopher who supported the Copernician theory of a heliocentric solar system. Burned at the stake by the Inquisition in Rome.

Cassini, Giovanni (1625-1712). Discoverer of four satellites and 4,800 km gap in ring system of Saturn.

Chandrasekhar, Subrahmanyam (b. 1910). Evolved theory of white dwarfs. Determined the mass limit for white dwarfs (Chandrasekhar limit). Won Nobel Prize (1983) for this work.

Copernicus, Nicholas (1473-1543). Presented sun-centered model of the solar system.

Descartes, René (1596-1650). Proposed first evolutionary cosmological theory.

Doppler, Christian (1803-1853). Discoverer of relationship between motion of an object and frequency of sound or light received and/or sent ("Doppler effect").

Drake, Frank (b. 1930). Leader of Project Ozma to detect intelligent extraterrestrial communications. Developed equation to estimate number of intelligent civilizations in the Galaxy.

Draper, Henry (1837-1882). American star cataloguer and astronomical photographer.

Eddington, Arthur (1882-1944). By observing the 1919 solar eclipse, proved that Einstein's prediction of the bending of starlight is valid. Explained the source of starlight as nuclear fusion; later verified by Bethe.

Einstein, Albert (1879-1955). Author of the special and general theories of relativity.

Flammarion, Camille (1842-1925). Researched the planet Mars. Popularized astronomy.

Flamsteed, John (1646-1719). First Astronomer Royal of Great Britain, 1675.

Foucault, J. Leon (1819-1868). Inventor of the Foucault pendulum, first physical proof that Earth rotates on its axis.

Fraunhofer, Joseph (1787-1826). Discovered absorption lines in the spectrum of the sun.

Galilei, Galileo (1564-1642). Italian astronomer who first used telescope to study the sky. Discovered craters and mountains on the moon, phases of Venus, and the four largest satellites orbiting Jupiter.

Galle, Johann (1812-1910). Observed Neptune for first time using position calculated by Leverrier.

Gamow, George (1904-1968). Astronomer who predicted the existence of the cosmic background radiation.

Glenn, John (b. 1921). First American astronaut to orbit Earth.

Goddard, Robert (1882-1945). Father of American rocketry, whose designs were the basis for today's modern rockets.

Hale, George Ellery (1868-1938). Astronomer most responsible for the construction of the 200″ Palomar telescope.

Hall, Asaph (1829-1907). Discovered the two moons of Mars — Phobos and Deimos.

Halley, Edmund (1656-1742). English astronomer. First to predict return of periodic comet, later named Halley's comet.

Hawking, Stephen (b. 1942). Proposed the existence of primordial black holes and described how they would evaporate and explode.

Henderson, Thomas (1798-1844). Measured first stellar parallax — Alpha Centauri, January 9, 1839.

Herschel, Caroline (1750-1843). First modern woman astronomer. Sister of William Herschel.

Herschel, John (1792-1871). Identified over 3,300 double stars and 500 nebulas. Son of William Herschel.

Herschel, William (1738-1822). English astronomer. Discoverer of the planet Uranus. First to develop modern model of Milky Way galaxy.

Hertzsprung, Ejnar (1873-1967). Cocreator of the Hertzsprung-Russell (temperature-luminosity) diagram.

Hevelius, Johannes (1611-1687). Star cataloguer, lunar mapmaker, constellation creator.

Hubble, Edwin (1889-1953). First to measure distance to Andromeda galaxy by using a cepheid variable. Discovered the velocity-distance relation — Hubble law.

Huggins, William (1824-1910). Discoverer of the stellar nature of the spiral nebula Andromeda (M31).

Huygens, Christian (1629-1695). Discoverer of Saturn's satellite Titan and Saturn's rings. Proposed wave theory of light.

Jansky, Karl (1905-1950). Discoverer of cosmic radio sources.

Kepler, Johannes (1571-1630). First to establish mathematical relationships to describe the motion of objects in the solar system (Kepler's three laws).

Kirchoff, Gustav (1824-1887). Formulated laws of spectroscopy.

Leavitt, Henrietta (1868-1921). Discovered period-luminosity relation for cepheid variables.

Lemaitre, George (1894-1966). Originated big bang theory of the origin of the universe.

Leverrier, Urbain Jean (1811-1877). French astronomer. Codiscoverer of Neptune.

Lowell, Percival (1855-1916). Builder of Lowell Observatory, dedicated to the study of Mars and the search for a planet beyond the orbit of Neptune.

Messier, Charles (1730-1817). Organized and established the famous Messier catalogue of nonstellar objects; for example: M31 — Andromeda galaxy.

Michelson, Albert (1852-1931). Proved that Earth's absolute velocity in space cannot be detected. Led Einstein to reject the ether.

Mitchell, Maria (1818-1889). First American woman astronomer.

Morley, Edward (1838-1923). Proved Earth's absolute velocity in space cannot be detected. Led Einstein to reject the ether.

Newton, Isaac (1642-1727). Inventor of the reflecting telescope. Discovered the law of universal gravitation and developed the type of mathematics known as calculus.

Olbers, Heinrich (1758-1840). Proposed "Olber's Paradox." Discovered numerous comets and asteroids.

Penzias, Arno (b. 1933). Codiscoverer of the cosmic background radiation.

Piazzi, Giuseppe (1746-1826). Discovered first asteroid, Ceres, January 1, 1801.

Ride, Sally (b. 1951). First American woman astronaut to fly in space.

Roemer, Olas (1644-1710). First person to measure accurately the speed of light.

Russell, Henry Morris (1857-1957). American astronomer — cocreator of the Hertzsprung-Russell (temperature-luminosity) diagram.

Schiaparelli, Giovanni (1835-1910). Discoverer of Martian "canali," later called canals. Named many features on Martian surface.

Schmidt, Maarten (b. 1929). Discoverer of the enormous redshifts in the spectra of quasars.

Schwabe, Heinrich (1789-1875). Discoverer of the eleven-year sunspot cycle.

Schwarzschild, Karl (1873-1916). First to describe the formation of the object now called a "black hole." The radius of the event horizon is called the Schwarzschild radius.

Shapley, Harlow (1885-1972). Established position of the sun within the Milky Way galaxy.

Shepard, Alan (b. 1923). First U.S. astronaut to fly in space (suborbital).

Tombaugh, Clyde (b. 1906). American astronomer who discovered Pluto.

von Braun, Wernher (1912-1977). Rocket expert. Led U.S. to landing a man on the moon.

Wheeler, John (b. 1911). Princeton physicist who first coined the phrase "black hole."

Wilson, Robert (b. 1936). Codiscover of the cosmic background radiation.

Glossary

Andromeda galaxy. Nearest major stellar system to the Milky Way. Located in the direction of Andromeda, it is spiral in shape and at a distance of 2.5 million light-years.

angstrom. A unit of length roughly the diameter of the hydrogen atom, 10^{-8} centimeter.

angular momentum. A measure of the momentum of a rotating object.

anisotropy of the CBR. Discovery of a slight non-uniform distribution of the CBR. The result of the Milky Way moving through space (Doppler effect).

Apollo. The United States' series of manned spacecraft missions to the lunar surface.

asteroid belt. Region of the solar system between Mars and Jupiter where most of the asteroids orbit the sun.

astronomical unit (AU). The average Earth-Sun distance — approximately 93 million miles or 140 million kilometers.

atmospheric windows. Wavelength regions of electromagnetic energy to which the atmosphere is "transparent."

atomic nucleus. The center of the atom containing the particles known as protons and neutrons.

Balmer lines. Spectrum of hydrogen in the visible range of electromagnetic energy. The result of transitions of electrons between the second and higher energy levels.

barycenter. The point around which each member of a binary system orbits. The center of mass of the binary system.

binary system. Two objects (the Earth-moon system or a stellar binary system).

bolometric magnitude. Brightness of an object determined by considering all forms of energy radiating from it: X-rays, UV, visible, infrared, etc. This measurement would be taken just outside Earth's atmosphere so no dimming effects could occur.

Callisto. Most remote of the Galilean satellites and most cratered, thus showing little erosion during its lifetime.

canals (Martian). Percival Lowell interpreted the "canali" seen by Schiaparelli as waterways or canals made by intelligent life.

cannibalism (between galaxies). Galaxies have been observed with two nuclei. Possibly the result of a slow-moving collision in which the members involved never separated.

Cassini's division. Easily seen gap between the brightest two rings of Saturn.

Chandrasekhar limit. Maximum mass for a white dwarf. Approximately 1.4 solar masses.

Chiron. Believed to be an asteroid — possibly a comet-like object that has an elongated orbit ranging from 8.5 to 20 astronomical units from the sun.

closed universe. A universe that will eventually stop expanding and begin collapsing until it comes together in a great catastrophe.

coronal hole. Dark regions in X-ray photographs of the corona. A region where very little X-ray energy is emitted. Believed to be the source of the solar wind.

cosmic jets. Jets of matter observed to have been ejected from the centers of some galaxies and quasars.

curvature of spacetime. Spacetime in the vicinity of a gravitational field. The mathematics of general relativity describe objects and energy moving along curved paths; hence the term curvature of spacetime.

Cygnus X-1. X-ray source in the constellation of Cygnus the swan. First object to be a candidate for a black hole.

declination. A coordinate for measuring the north-south positions of celestial bodies. The angle measured northward or southward from the celestial equator to the object.

Deimos. The smaller of the two moons of Mars.

density wave. One explanation for the maintenance of the spiral arms. Wave crest of matter, emanating from the center of the Galaxy, moving through the disk. Causes regions of higher density enhancing star formation.

differential rotation. Describes the fact that the sun and probably other stars do not rotate as a solid unit. The equatorial region of the sun rotates more rapidly than the polar regions.

dispersion. The spreading of white light into the colors of the spectrum (red, orange, yellow, green, blue, indigo, violet).

Doppler effect. The phenomenon explaining the observed change in frequency of sound or radiation as a result of the relative motion, along the line of sight, between the source and the observer. (Red shift — receding; blue shift — approaching).

$E = mc^2$. Equivalence of matter and energy equation. Nuclear energy equation. E = energy, m = mass, c = speed of light.

electron degeneracy. In a white dwarf, gravitational pressure has compressed electrons so close together that they cannot be compressed any further. The forces between electrons prevent the star from collapsing any further.

ellipse. The geometrical shape observed to be the orbits of the planets. An "oval" with defined characteristics.

emmision spectrum. Spectrum consisting of several discrete bright lines seen against a dark background.

energy levels (orbits). According to the Bohr model of the atom, circular orbits in which electrons move around the nucleus of the atom.

ergosphere. Region around a rotating black hole in which one cannot remain at rest.

Europa. Galilean satellite next out from Io, observed to have lines crisscrossing the surface. May have ice sheet (glacier) a mile thick covering the surface.

exosphere. Outermost layer of Earth's atmosphere.

F ring. The twisted or braided ring of Saturn that appears to be under the control of the shepherding satellites.

flash spectrum. Bright line spectrum of the sun's chromosphere taken just prior to or just after the instant of totality.

flat universe. Flat geometry (Euclidean) of universe that expands far enough so that the universe stops expanding at an infinite future time and so does not collapse together.

Foucault pendulum. A pendulum devised in 1851 to demonstrate that Earth rotates on its axis. The pendulum ball was set to oscillate in a single plane. It was soon observed that the plane was changing. Since no force was acting to change the plane of swing, it had to be concluded that the building and hence Earth were turning underneath the plane of oscillation.

Fraunhofer lines. Absorption lines in the spectrum of the sun that were catalogued by Joseph Fraunhofer.

fundamental forces. Scientists believe there exists in the universe only four forces: gravity, (it holds us on to Earth), electromagnetic (this force bonds the elections in atoms and holds atoms together to form compounds and other substances), strong nuclear force (it bonds the protons and neutrons into the nucleus of the atom), and weak nuclear force (it plays a role in radioactive decay).

Galilean satellites (of Jupiter). The four largest satellites of Jupiter, all discovered by Galileo.

Ganymede. Largest of the Galilean satellites and next out from Europa. Largest satellite in the solar system.

geocentric. Earth-centered.

gravity assist. The use of the gravity of one planet to propel the spacecraft on to another planet. Mariner 10 used Venus to propel it to Mercury and the Pioneer and Voyager craft used Jupiter to propel them on to Saturn.

Great Red Spot. Large red oval seen in the clouds of Jupiter.

helium flash. The beginning of helium fusion (the triple alpha process) occurring in the dense cores of red giant stars.

highlands (lunar). The brighter regions on the surface of the moon. These regions show more cratering and have more mountains than the maria and are therefore older than the maria.

Hubble classification. System, developed by Edwin Hubble, that classifies galaxies according to their shape and content.

Hubble law. The greater the distance to the galaxy the greater the velocity of recession. A direct proportion — double the distance and the velocity of recession is doubled.

infrared radiation. Electromagnetic radiation with wavelengths longer than red light and shorter than radio waves. Also called heat radiation.

intelligent life equation. Equation developed by Frank Drake that tries to determine the number of intelligent civilizations in the Galaxy.

interferometry. Technique used to measure and resolve extremely small celestial sources with great accuracy. Using optical telescopes meters apart and radio telescopes miles apart, the technique employs the interference pattern created when the energy from the object reaches each telescope at a different time to resolve very fine detail in the celestial source.

inverse-square law of light. Relationship between intensity of the radiation and distance: If the distance is double, the intensity appears one-fourth as great; if the distance becomes one-third as great, the brightness would be nine times greater.

Io. Galilean satellite nearest to Jupiter. Observed to have active volcanoes on its surface.

Jovian planets. Large liquid-gaseous planets, made mostly of hydrogen and helium with no solid surface. Similar to the planet Juptier.

length contradiction. The shortening of length in the direction of motion of an object relative to a non-moving reference frame.

liquid metallic hydrogen. Hydrogen under such extremely high pressure that the element takes on the properties of a liquid and metal, i.e., it conducts electricity.

liquid molecular hydrogen. Hydrogen under sufficiently high pressure so that the element, normally a gas, takes on the properties of a liquid.

LMC X-3. X-ray source in the large Magellanic Cloud. Likely candidate for a black hole.

luminosity function. The relative number of stars in successive intervals of absolute magnitude within any given volume of space.

lunar orbiter. United States spacecraft that orbited the moon. Photographed the near and far sides of the moon in great detail in preparation for the Apollo lunar landing.

Lunokhod. Soviet spacecraft that soft-landed on the lunar surface and then deployed a rover to explore the nearby region.

Mariner 6, 7. Two spacecraft that flew by Mars and photographed its surface. Came within 3,500 km of the planet.

Mariner 9. Spacecraft that orbited Mars and took thousands of photographs. Led to the discovery of huge volcanoes, great canyons, and ancient "water" channels.

mass increase (with velocity). Increase in mass of a moving object relative to a frame of reference that is not in motion.

microwave radiation. Electromagnetic radiation with wavelengths and energy between that of infrared and radio wavelengths — short wave radio energy.

missing mass. Amount of matter still needed to bring the expanding universe to a halt and begin the collapse.

negatively curved universe. Saddle-shaped geometry of an open universe.

neutrino mass. Particle originally thought to have zero rest mass. May have some small mass that is enough to close the universe.

neutrino "telescope." Device designed to detect neutrinos.

nodes. The points of intersection of the plane of the orbit of a celestial object (such as the moon's orbit about Earth) and the plane of the ecliptic.

nova. Rapid brightening of a star to many hundreds or thousands of times its original brightness. Usually associated with a binary system, in which one of the companions is a white dwarf. May occur more than once.

nuclear bulge. Central concentration of stars and gas at the center of a galaxy. The nucleus of a galaxy.

nuclear reaction. Reaction affecting the nucleus of the atom either by splitting it into smaller pieces (fission) or combining it with other nuclei (fusion). Both reactions yield great amounts of energy.

obliquity. The angle between the planes of the equator of a planet and the ecliptic.

Olympus Mons. Largest volcano on surface of Mars.

Oort cloud. Cloud of comets surrounding the sun at a distance of some 50,000 astronomical units. Perturbations caused by nearby stars cause the comet to fall into the solar system.

open universe. Describes the universe expanding forever.

pair production. The conversion of pure energy into a pair of particles, such as an electron and positron — observed regularly in nuclear physics laboratories.

peculiar galaxy. Galaxy with unusual characteristics such as shape, energy output, or spectrum.

Phobos. The larger of the two moons of Mars.

photon sphere. Sphere of orbiting light beams around a black hole.

polar ice caps (Martian). Composed mostly of frozen carbon dioxide. There is a residual cap of water ice. The caps change in size with the seasons of Mars.

polarization of starlight. As starlight passes through the interstellar medium, the material within the interstellar medium causes the transverse component of the electromagnetic radiation to be confined to a single plane (plane polarized) or to one that rotates (circularly polarized light).

positively curved universe. Spherical-shaped geometry of a closed universe.

precession. The wobble of Earth on its axis, the result of the gravitational influences of the sun and moon. The most noticeable effect is the change in the North Star. At the present, Earth's polar axis is pointed toward the star Polaris, which is hence called the North Star. In 14,000 A.D., precession will cause the axis to move and point toward the star Vega, in the constellation of Lyra the Harp. Five thousand years ago, the "North Star" was Thuban, in Draco the Dragon. It takes 26,000 years for the cycle to be repeated. In other words, Polaris will be the North Star again in 28,000 A.D.

primordial black hole. Mini black hole — size of an electron or smaller, predicted by Stephen Hawking to have been formed at the time of the big bang. May also evaporate and explode.

Problicom. Projection blink comparator. Device used to discover objects that are moving through space or changing brightness. Invented by Ben Mayer.

Project Ozma. First concerted effort to listen for intelligent communications from extraterrestrial civilizations.

Project SETI. Search for extraterrestrial intelligence. Project in operation during the 1980s.

proper motion. The angular speed with which a star seems to move across the sky perpendicular to our line of sight. Usually expressed in seconds of arc per year.

PSR 1913 + 16. Pulsar in a binary system with another neutron star or white dwarf. Research shows that this object is emitting gravitational radiation as predicted by the theory of general relativity.

radar observations. The use of reflected radio signals to describe features on the surface of a planet. The rotation rate of Mercury was determined in this way.

radio wave. Longest wavelength and lowest energy form of electromagnetic radiation.

Ranger 7, 8, 9. Series of United States spacecraft that crash landed on the moon and sent back several thousand photographs.

ringlets. Thin rings that make up the larger and brighter rings of Saturn. Give the rings of Saturn a groove-like appearance similar to a phonograph record.

rotating black holes. Black holes that are spinning. The solution of the general theory of relativity field equations, discovered by Roy P. Kerr; includes angular momentum. Sometimes called a Kerr black holes.

Sagittarius A. Very bright radio source in the direction of the center of the Galaxy. Believed to originate at the center of the Galaxy.

Saturn's rings. Disk of thousands of tiny particles ranging in size from grains of sand to larger rocks in orbit about the planet.

Schmidt telescope. A type of reflecting telescope that enables the astronomer to take wide angle photographs without suffering from the aberrations normally produced by a spherical concave mirror.

seasons (Martian). Refers to the changing color patterns seen on the surface of Mars and the changing size of the polar ice caps.

shepherding satellites. Small satellites or moons in orbit around Saturn that appear to control some of the structure in the rings of Saturn.

singularity. The center of a black hole. Defined as a region of infinite curvature of spacetime.

solar wind. Charged particles escaping from the sun.

spectroheliograph. An instrument for photographing the sun or part of the sun in the single color (monochromatic) light of a particular spectral line.

speed of light. Velocity at which light travels, 186,282 miles per second or 300,000 kilometers per second. According to the principle of relativity, this velocity is constant and the same for all observers regardless of their relative motions.

spokes. Dark lines in the B ring of Saturn that extend radially outward from the planet.

SS 433. A neutron star in a binary system in which the larger companion dumps matter onto the neutron star. The results are unusual X-ray radiation and two jets of matter pointed in opposite directions.

star gauging. Procedure used by William Herschel to count the stars seen through his telescope. This procedure eventually led Herschel to certain conclusions about the shape of the Galaxy.

strong nuclear force. Force that binds the protons and neutrons in the nucleus of the atom. It is significant only over very short distances as "nuclear glue." One hundred times as strong as the electromagnetic force.

superclusters. Large numbers of galaxies and groups of galaxies that appear to cluster together. Their distribution reveals regions of space that seem to be devoid of galaxies.

superluminal velocity. Faster than the speed of light. An illusion observed in some components of some quasars.

Surveyor. Series of United States spacecraft that soft-landed on the lunar surface.

synchronous. When a planet's period of rotation or revolution is a whole number multiple of another period of rotation or revolution. For example, Mercury orbits the sun twice as it spins on its axis three times.

terrestrial planets. Planets similar to Earth with a hard, rocky, solid surface, and relatively small in size.

time dilation. Effect predicted by special relativity. The slowing of time of a moving clock relative to a stationary clock on Earth.

Titan. Largest moon of Saturn and second largest in the solar system. Contains a dense atmosphere of nitrogen.

UFO. Unidentified flying object.

universal gravitation. The tendency of matter to attract itself. The word "universal" implies that the equations derived on Earth to describe gravity apply throughout the universe.

Valles Marineris. Valley of the Mariners. Large canyon on the surface of Mars.

vernal equinox. The position of the sun on or about March 22. The first day of spring. The point on the celestial sphere that is the origin for the measurement of the celestial coordinate, the right ascension.

Viking lander. Spacecraft that soft-landed on Mars. Conducted three experiments designed to detect life.

Viking orbiter. Part of a two section spacecraft sent to Mars that remained in orbit while the other section landed on the surface. Made detailed photographic mapping of the surface.

wavelength. The distance between one wave crest or trough and another.

weak nuclear force. Force associated with radioactivity and the decay of neutrons into protons and electrons; a million times weaker than the electromagnetic force.

zenith. The point in the sky that is directly above the observer.

Bibliography

Lesson 1

Gore, Richard. 1983. The once and future universe. *National Geographic* 163, no. 6 (June): 704.

Horodyski, J.M. 1986. Reach for the stars: The story of Mt. Wilson Observatory. *Astronomy* 14 (December): 6-19ff.

Lovi, G. 1985. Tangled sky names. *Sky and Telescope.* 65 (April): 335-337.

Phillips, Henry, and Robert Burnham. 1977-1978. Learning the constellations, parts 1-4. *Astronomy* 5 & 6: 42, 36, 46.

Reed, G. 1986. Astronomy and astrology. East vs. west. *Astronomy* 14 (January): 90-95.

Skylore [Comets]. 1985. *Astronomy.* 13 (June): 47.

Lesson 2

Aveni, A.F. 1986. Archaeoastronomy: Past, present, future. *Sky and Telescope* 72 (November): 456-460.

Deitmar, S. 1985. The Observatories of Jai Singh. *Astronomy* 13 (January): 18-22.

Drake, Stillman. 1980. Newton's apple and Galileo's dialogue. *Scientific American* (August): 150.

Gingerich, Owen. 1982. The Galileo affair. *Scientific American* (August): 132.

Gingerich, Owen. 1984. Galileo and the phases of Venus. *Sky and Telescope* 68 (December): 520-522.

History of American astronomy. 1976. *Astronomy* (July). (Note: the entire issue is devoted to American astronomers.)

McMullin, E. 1987. Bruno and Copernicus. *Isis* 78 (March): 55-74

Lesson 3

Berry, Richard. 1978. Photograph the moon. *Astronomy* (February): 34.

Eclipse blindness: A casualty report. 1985. *Sky and Telescope* 69 (April): 315.

Espenak, Fred. 1980. North American solar eclipses — the next fifty years. *Astronomy* (November).

The moon — The Southern Highlands; Return to the moon; The Moon — Ring mountains and walled plain: The Moon — Rilles and wrinkle ridges. *Astronomy*. June 1979, April 1980, December 1980, May 1981, respectively.

The Moon's brighest crater. 1986. *Sky and Telescope* 72 (September): 268-269.

Rotton, J., and I.W. Kelly. 1986. The Lunacy of it all: Lunar phases and human behavior. *Mercury* 15 (May/June): 73-75ff.

Stephenson, F. Richard. 1982. Historical eclipses. *Scientific American* (October).

Whiteman, M. 1986. Eclipse predictions on your computer. *Astronomy* 14 (November): 67-70

Lesson 4

Augensen, Harry, and Jonathan Woodbury. 1982. The electromagnetic spectrum. *Astronomy* (June).

Darling, D.J. 1984. Spectral visions: The long wavelengths. *Astronomy* 12 (August): 16-22.

Darling, D.J. 1984. Spectral visions: The short wavelengths and beyond. *Astronomy* 12 (September): 14-22.

Davies, John K. 1987. The extreme ultraviolet: A promising new window on the universe. *Astronomy* 15 (July): 82-87.

Great interstellar atoms. 1985. *Sky and Telescope* 70 (December): 547-548.

Parker, B. 1986. The mysterious cosmic x-ray background. *Astronomy* 14 (February): 90-95.

Lesson 5

Gordon, M.A. 1985. VLBA — A continent size radio telescope. *Sky and Telescope* 69 (June): 487-490.

Hanle, P.A. 1985. Astronomers' Congress and the large space telescope. *Sky and Telescope* 69 (April): 300-305.

The IRAS harvest. 1985. *Sky and Telescope* 70 (August): 107-108.

Mammana, Dennis. 1985. Through monster eyes. *Astronomy* 13 (February): 6-17.

Minton, R.B. 1985. Spectral lines. *Sky and Telescope* 69 (June): 545-546.

Shore, L.A. 1987. IUE: Nine years of astronomy. *Astronomy* 15 (April): 14-22.

Shore, L.A. 1987. The telescope that never sleeps. *Astronomy* 15 (August): 14-22.

Stencel, Robert, William Blair, and Susan Conat-Stencel. 1978. Astronomical spectroscopy. *Astronomy* (June): 6.

Lesson 6

Clark, David H. 1980. Extraterrestrial climate threats. *Astronomy* 8, no. 8 (August): 66.

Cordell, B.M. 1986. Mars, Earth, and ice. *Sky and Telescope* 72 (July): 117-22.

Gillett, Stephen L. 1985. The rise and fall of the early reducing atmosphere. *Astronomy* 13 (July): 66-71.

Grasshoff, R. 1985. Comets, "death stars," and mass extinctions: The rise and fall of a scientific theory. *Astronomy* 13 (May): 18-22.

Payne, M.M. 1987. Does the earth rotate? *Physics Teacher* 25 (February): 86-87.

Rabaiotti, A.G. 1985. Spacelab. Earth observations. *Astronomy* 13 (May): 58.

Toon, O.B., and S. Olson. 1985. The warm earth. Our planet didn't broil. It didn't freeze. What has kept it liveable. *Science 85* 6 (October): 50-57.

Lesson 7

Boss, A.P. 1986. The origin of the moon. *Science* 231 (24 January): 341-345.

Brownlee, S. 1985. A whacky theory of the moon's birth. *Discover* 6 (March): 64-69.

Kitt, M.T. 1987. Sculpting the moon. *Astronomy* 15 (February): 82-87.

Register, B.M. 1985. The fate of the moon rocks. *Astronomy* 13 (December): 14-22.

Tektites and lunar volcanoes. 1987. *Sky and Telescope* 73 (May): 481.

Lesson 8

Cordell, B.M. 1984. Mercury: The world closest to the sun. *Mercury* 13 (September/October): 136-146.

Hall, L. 1987. Geology of Venus. *Physics Today* 40 (January): S37-S38.

James, Warren. 1982. Unveiling Venus with VOIR. *Sky and Telescope* (February): 141.

Kobrick, Michael. 1982. Topography of the terrestrial planets. *Astronomy* 10, no. 5 (May): 18.

Maran, S.P. 1985. Very volcanic Venus. *Natural History* 94 (November): 34ff.

Mercury's atmosphere is rich in sodium. 1985. *Sky and Telescope* 70 (November): 430-431.

Prinn, R.G. 1985. The volcanoes and clouds of Venus. *Scientific American* 252 (March): 46-53.

Shawcross, W.E. 1985. Venus and the Maya. *Sky and Telescope* 70 (August: 111-114.

Lesson 9

Carroll, Michael. 1987. The changing face of Mars. *Astronomy* 15 (March): 6-22.

Cordell, Bruce. 1983. The first Martians. *Astronomy* (March): 6.

Haberle, R.M. 1986. The climate of Mars. *Scientific American* 254 (May): 54-62ff.

Kanipe, J. 1986. Springtime on Mars: A rare opposition. *Astronomy* 14 (June): 80-85.

Lowell, Percival. 1895. *Mars*. Boston: Houghton Mifflin & Co.

Oberg, James. 1978. Terraforming. *Astronomy* (May): 6.

Strickland, Edwin. 1980. The colors of Mars. *Astronomy* (May): 6.

Toon, O.B., and S. Olson. 1985. How Mars froze . . . and Venus broiled. *Science 85* 6 (October): 56-57.

Wallace, Alfred Russel. 1907. *Is Mars inhabitable?* London: Macmillan & Co., Ltd.

Lesson 10

Bennett, G.L. 1987. Return to Jupiter (Galileo Probe). *Astronomy* 15 (January): 6-15.

Carroll, M.W. 1987. Project Galileo: The phoenix rises. *Sky and Telescope* 73 (April): 359-361.

Elliott, J., and R. Kerr. 1985. How Jupiter's ring was discovered. *Mercury* 14 (November/December): 162-171.

Schwartzenburg, Dewey. 1980. The great red spot. *Astronomy* (July): 6.

Simon, S. 1986. The view from Europa. *Astronomy* 14 (November): 98-103.

A truly extraterrestrial UFO. 1987. *Sky and Telescope* 73 (June): 603-604.

Lesson 11

Esposito, Larry W. 1987. The changing shape of planetary rings. *Astronomy* 15 (September): 6-17.

Morrison, N.D., and S. Gregory. 1985. The exotic atmosphere of Titan. *Mercury* 14 (September/October): 154-157.

Osterbrock, D.E. 1985. The nature of Saturn's rings. *Mercury* 14 (March/April): 46-47ff.

Owen, Tobias. 1982. Titan. *Scientific American* (February): 98.

Robertson, D.F. 1987. Cassini. *Astronomy* 15 (September): 21-24.

Lesson 12

Bennett, J.A. 1981. The discovery of Uranus. *Sky and Telescope* (February): 188.

Berry, Richard. 1986. Voyager: Discovery at Uranus. *Astronomy* 14 (May): 6-22.

Brown, R.H., and D.P. Cruikshank. 1985. Moons of Uranus, Neptune, and Pluto. (ground-based observations) *Scientific American* 253 (July): 38-47.

Croswell, K. 1986. Pluto enigma on the edge of the solar system. *Astronomy* 14 (July): 6-10ff.

Grosser, Morton. 1979. *Discovery of Neptune.* New York: Dover Publications, Inc.

Johnson, T.V. 1987. The moons of Uranus [Voyager 2]. *Scientific American* 256. (April): 48-60.

Kerr, R.A. 1986. Voyager finds Uranian shepherds and a well-behaved flock of rings. *Science* 23 (21 February): 793-796.

Kohlhase, C.E. 1986. On course for Neptune. *Astronomy* 14 (November): 6-15.

Levy, D.H. 1986. Clyde Tombaugh is 80; A look back at Pluto. *Astronomy* 14 (February): 26ff.

Neptune's "ring" may be an incomplete arc. 1986. *Astronomy* 14 (March): 79-80.

O'Meara, S.J. 1985. A visual history of Uranus. *Sky and Telescope* 70 (November): 411-414.

Smith, B.A. 1986. Voyager 2 in the Uranian system: Imaging science results. *Science* 233 (4 July): 43-64.

Tombaugh, C.W. 1986. The discovery of Pluto: Some generally unknown aspects of the story. *Mercury* 15 (May/June): 66-72.

Lesson 13

Alvarez, Luis. 1987. Mass extinctions caused by large bolide impacts. *Physics Today* 40 (July): 24-33.

Berry, Richard. 1987. Search for the primitive. *Astronomy* 15 (June): 6-22.

Calder, Nigel. 1980. *The comet is coming.* New York: Viking Press.

Davis, Joel. 1980. Asteroids — The flying mountains. *Astronomy* 8, no. 5 (May): 16.

Davis, Joel. 1981. The tektite controversy. *Astronomy* 9, no. 4 (April): 6.

Gingerich, Owen. 1986. Newton, Halley, and the comet. *Sky and Telescope* 14 (March): 63-66.

Greenstein, G. 1985. Heavenly fire. [1908 explosion in Tunguska area of Siberia: What caused it?] *Science 85* 6 (July/August): 70-77.

Kerr, R.A. 1985. Antarctic meteorites: Are they a breed apart? *Science* 228 (28 June): 1520.

Meteorites escape from Mars. 1987. *Astronomy* 15 (February): 78.

Powell, B.E. 1986. A Halley scrapbook: What we did in 1910. *Astronomy* 14 (March): 24ff.

Whipple, F.L. 1987. The black heart of Comet Halley. *Sky and Telescope* 73 (March): 242-245. (Note: A number of other articles covering the 1985/86 apparition of the comet are included in this issue.)

Lesson 14

Bennet, G.L. 1987. Voyage into the third dimension [Ulysses Mission] *Astronomy* 15 (May): 14-22.

Golub, Leon. 1982. What heats the solar corona? *Astronomy* (September): 74.

Lo Presto, J.C. 1986. The rotation of the sun. *Astronomy* 14 (December): 106-111.

Maran, S.P. 1984. Flare-ups on the sun. *Natural History* 93 (April): 90-95.

Stahl, Philip A. 1980. Sunspots. *Astronomy* (May): 66.

Williams, G.E. 1986. The solar cycle in Precambrian time. *Scientific American* 255 (August): 88-96.

Wolfson, Richard. 1983. The active solar corona. *Scientific American* (February).

Lesson 15

Bahcall, J. 1987. Solutions to the solar neutrino problem. *Physics Today* 40 (January): S11.

Franco, A., and D.H. Smith. 1987. Vanishing solar neutrinos. *Sky and Telescope* 73 (February): 149.

Learned, John G. and David Eichler. 1981. A deep sea neutrino telescope. *Scientific American* (February): 138.

Missing neutrinos and missing mass. 1985. *Sky and Telescope* 70 (November): 431-432.

Mullan, D.J. 1984. Tuning in to the interior of a star. *Astronomy* 12 (December): 66-70.

Schramm, David N., and Robert N. Clayton. 1978. Did a supernova trigger the formation of the solar system? *Scientific American* (October): 124.

Lesson 16

Croswell, K. 1987. Visit the nearest stars. *Astronomy* 15 (January): 16-22.

Lovi, G. 1986. Our galactic suburb. *Sky and Telescope* 72 (November): 491-492.

Malin, D. 1986. In search of star colors. *Sky and Telescope* 72 (October): 326-330.

Rabaiotti, A.G. 1986. Hipparcus satellite to recatalog the stars. *Astronomy* 14 (February): 79-80.

Reddy, Francis. 1983. How far the stars. *Astronomy* (June): 6.

Witkoski, F.M. 1985. Fourteen special stars. *Astronomy* 13 (April): 75-77.

Lesson 17

Kaler, J.B. 1986. Cousins of our sun: The G stars. *Sky and Telescope* 72 (November): 450-453.

Kaler, J.B. 1986. M stars: Supergiants to dwarfs. *Sky and Telescope* 71 (May): 450-454.

Kaler, J.B. 1986. Origins of the spectral sequence. *Sky and Telescope* 71 (February): 129-134.

Kaler, J.B. 1987. The temperate F stars. *Sky and Telescope* 73 (February): 131-134.

Lesson 18

Finley, David. 1981. The search for extra solar planets. *Astronomy* 9, no. 12 (December): 90.

Hopkins, J.L., and R.E. Stencel. 1986. Epsilon Aurigae. *Astronomy* 14 (February): 6-11.

MacRobert, A. 1986. Optical double stars. *Sky and Telescope* 71 (March): 280-281.

Morrison, N.D., and S. Gregory. 1986. The puzzle of Epsilon Aurigae. *Mercury* 15 (November/December): 177-180.

Paczynski, B. 1984. Binary stars. *Science* 225 (20 July): 275-280.

Patton, Paul. 1982. Three suns of Centaurus. *Astronomy* 10, no. 1 (January): 6.

Lesson 19

Blitz, Leo. 1982. Giant molecular cloud complexes in the Galaxy. *Scientific American* (April): 84.

Clark, G.O. 1985. Ancients of the universe. The globular clusters. *Astronomy* 13 (May): 6-13.

Clark, G.O. 1986. Stellar populations: Key to the clusters. *Astronomy* 14 (October): 106-111.

Gingerich, Owen. 1984. Discovery of the Milky Way's spiral arms. *Sky and Telescope* 68 (July): 10-12.

Havlen, R.J. 1987. New determination of the distance to the galactic center. *Physics Today* 40 (January): S7-S8.

Hertz. P.L. 1985. X-ray sources in globular clusters and in the plane of the galaxy. *Mercury* 14 (March/April): 42-43.

Marschall, Laurence. 1982. Secrets of interstellar clouds. *Astronomy* 10, no. 3 (October): 6.

Odenwald, S. 1985. The road to discovery. *Astronomy* 13 (August): 16-22.

Smith, D.H. 1985. Reflection nebulae, celestial veils. *Sky and Telescope* 70 (September): 207-210.

Tucker, Wallace. 1981. The star clouds of Magellan. *Astronomy* 9, no. 10. (October): 6.

Lesson 20

Crossen, Craig. 1983. Studying galactic structure with binoculars. *Astronomy* 11, no. 7. (July): 50.

Marschall, Laurence. 1982. Galactic coronas. *Astronomy* 10, no. 11 (November): 74.

Milky Way: Site of radio threads and jets. 1986. *Astronomy* 14 (May): 70-71.

Morrison, N.D., and S. Gregory. 1986. The nucleus and nuclear bulge of our galaxy. *Mercury* 14 (July/August): 119-121.

Mulholland, D. 1985. The beast at the center of the galaxy. *Science 85* 6 (September): 50-55.

Smith, D.H., and L.J. Robinson. 1984. Dissecting the hub of our galaxy. *Sky and Telescope* 68 (December): 494-497.

Lesson 21

Beatty, J.K. 1985. Is anyone there? *Science 85* 6 (June): 34-37

Clarke, A.C. 1985. Counting down for a flight to the stars. *Discover* 6 (July): 54-58

Darling, David. 1987. Star trek — Part one. The adventure begins. *Astronomy* 15 (March): 94-99.

Darling, David. 1987. Star trek — Part two: To distant shores. *Astronomy* 15 (April): 94-99

Did Martian life form in ice-covered lakes? 1987. *Astronomy* 15 (March): 77-78.

Olson, E.C. 1985. Intelligent life in space. *Astronomy* 13 (March): 77-78.

Parker, Barry. 1979. Are we the only intelligent life on our galaxy? *Astronomy* 7, no. 1 (January): 6.

Ready, SETI, Go! 1983. *Discover* 4 (May): 86

Sagan, Carl. 1973. *The cosmic connection.* Garden City, New York: Anchor Press-Doubleday.

Sagan, Carl, and Frank Drake. 1975. The search for extraterrestrial intelligence. *Scientific American* (May).

Shklovskii, I.S., and Carl Sagan. 1966. *Intelligent life in the universe.* New York: Dell Publications.

A telescope to search for earthlike planets. 1986. *Astronomy* 14 (December): 83-84

Lesson 22

Abbot, Edwin. 1952. *Flatland* New York: Dover Press.

Barnett, Lincoln. 1957. *The universe and Dr. Einstein.* New York: Sloane Publishing.

Calder, Nigel. 1979. *Einstein's universe.* New York: Viking Press.

Gardner, Martin. 1962. *Relativity for the millions.* New York: Macmillan Publishing.

Haugan, M.P., and C.M. Will. 1987. Modern tests of special relativity. *Physics Today* 40 (May): 69-76. (Note: There are several articles on relativity in this issue; however, these articles are more detailed and sophisticated than the average.)

Henry, R.C. 1985. Special relativity made transparent. *Physics Teacher* 23 (December): 536-539.

Pinheiro, Robert. 1979. Gravity wave astronomy. *Astronomy* 7, no. 6. (June): 6.

Roman, T.A. 1986. General relativity, black holes, and cosmology: A course for nonscientists. *American Journal of Physics* 54 (February): 144-151.

Lesson 23

Boss, A.P. 1985. Collapse and formation of stars. *Scientific American* 252 (January): 40-45.

Croswell, K. 1984. Stars too small to burn. *Astronomy* 12 (April): 16-22

Croswell, K. 1985. When stars coalesce. *Astronomy* 13 (May): 67-69ff.

Fortier, E. 1987. Touring the stellar cycle. *Astronomy* 15 (March): 49-53.

Fowler, W.A. 1984. The quest for the origin of the elements. *Science* 226 (23 November): 922-935.

Lada, Charles J. 1982. Energetic outflows from young stars. *Scientific American* (July): 82.

Mullan, D.J. 1984. Tuning in to the interior of a star. *Astronomy* 12 (December): 66-70.

Nemiroff, R.J. 1986. The main sequence blues. *Astronomy* 14 (January): 24.

Lesson 24

Darling, David. 1985. Breezes, bangs, and blowouts: Stellar evolution through mass loss. *Astronomy* 13 (September): 78-83; (November): 94-98.

Levy, D.H., and P. Jedicke. 1987. Betelgeuse. *Astronomy* 15 (April): 6-13.

Nicastro, A.J.. 1984. White dwarfs: Big things in small packages. *Astronomy* 12 (July): 6-14.

Parker, B. 1984. Those amazing white dwarfs. *Astronomy* 12 (July): 15-22.

Shipman, H. 1987. Some record-breaking white dwarf stars. *Physics Today* 40 (January): S8-S9.

The strange case of Zeta Aurigae. 1983. *Astronomy* (March).

Trimble, V. 1986. White dwarfs: The once and future suns. *Sky and Telescope* 72 (October): 348-353.

Lesson 25

Bethe, H.A., and G. Brown. 1985. How a supernova explodes. *Scientific American* 252 (May): 60-68.

Davis, Joel. 1980. The puzzle of SS 433. *Astronomy* 8, no. 7. (July): 66.

Greenstein, G. 1985. Neutron stars and the discovery of pulsars. *Mercury* 14 (March/April): 34-39ff; (May/June): 66-73.

Pennypacker, Carl. 1987. Searching for supernovae. The discovery in M-99. *Astronomy* 15 (August): 75-79.

Richardson, Steven M., and Lee Anne Willson. 1980. Cosmic contamination. *Astronomy* 8, no. 6 (June): 6.

Schorn, R.A. 1987. Supernova shines on. *Sky and Telescope* 73 (May): 470-475.

Shaham, J. 1987. The oldest pulsars in the universe. *Scientific American* 256 (February): 50-56.

Supernova [1987A] 1987. *Scientific American* 256 (May): 54ff.

Talcott, Richard. 1987. A burst of discovery: The first days of Supernova 1987A. *Astronomy* 15 (June): 90-95.

Weisberg, Joel M., Joseph H. Taylor, and Lee A. Fowler. 1981. Gravitational waves from an orbiting pulsar. *Scientific American* (October): 74.

Lesson 26

Darling, David. 1983. Quest for black holes. *Astronomy* 11 (July): 6-22.

Golden, Frederic. 1976. *Quasars, pulsars and black holes.* New York: Charles Scribner's Sons.

Hall, D.E. 1986. The hazards of encountering a black hole: A variety of ways to die. *Physics Teacher* 24 (January): 29-33.

Hutchings, J.B. 1984. LMC X-3 — A black hole in a neighbor galaxy. *Mercury* 13 (July/August): 106-107.

Parker, B. 1986. In and around black holes. *Astronomy* 14 (October): 6-15.

Shipman, Harry L. 1976. *Black holes, quasars, and the universe.* Boston: Houghton Mifflin.

Lesson 27

Finkbeiner, A.K. 1985. Cold dark matter and the origin of galaxies. *Astronomy* 13 (April): 67-71.

Hodge, P. 1987. The local group: Our galactic neighborhood. *Mercury* 16 (January/February): 2-15.

Parker, B. 1985. Celestial pinwheels: The spiral galaxies. *Astronomy* 13 (May): 14-22.

Rabalais, M. 1986. How to conquer the Messier objects. *Astronomy* 14 (December): 88-91. (Gives observation techniques.)

Reddy, Francis. 1983. To sculpt the galaxies. *Astronomy* 11, no. 1 (January): 6.

Silk, J. 1986. Formation of the galaxies. *Sky and Telescope* 72 (December): 582-587.

Smith, D.H. 1986. A sideways look at galactic history. *Sky and Telescope* 72 (August): 111-114.

Tucker, Wallace. 1981. The star clouds of Magellan. *Astronomy* 9, no. 10 (October): 6.

Lesson 28

Blandford, Roger, Mitchell Bezelman, and Martin Rees. 1982. Cosmic jets. *Scientific American* (May): 116.

Chaffee, Jr., Frederick H. 1980. Discovery of a gravitational lens. *Scientific American* (November): 70.

Gregory, S., and N.D. Morrison. 1986. A quasar/galaxy double image. *Mercury* 15 (January/February): 25-27.

Kanipe, J. 1987. M-87: Describing the indescribable. *Astronomy* 15 (May): 6-13.

Nearest gravitational lens. 1985. *Sky and Telescope* 69 (April): 315-316.

Quasars faster than the speed of light? 1987. *Astronomy* 15 (February): 79ff.

Schendel, Jack. 1982. Looking inside quasars. *Astronomy* 10, no. 11 (November): 6.

Shaffer, David B., and Gregory A. Shields. 1980. Why all the fuss about quasars? *Astronomy* 8, no. 10 (October): 6.

Shipman, Harry L. 1976. *Black holes, quasars, and the universe.* Boston: Houghton Mifflin.

Lesson 29

Bauer, Claude. 1986. COBE's (Cosmic Background Explorer) quest. *Astronomy* 14 (August): 74-79.

Corwin, Mike, and Dale Wachowiak. 1985. Discovering the expanding universe. *Astronomy* 13 (February): 18-22.

Ferris, T. 1987. Where are we going? *Sky and Telescope* 73 (May): 486-490.

Gregory, Stephen, and Laird Thompson. 1982. Superclusters and voids in the distribution of galaxies. *Scientific American* (March): 106.

Kippenhahn, R. 1987. Light from the depths of time. *Sky and Telescope* 73 (February): 140-142.

Melott, Adrian. 1983. Cosmology on a computer, parts 1 & 2. *Astronomy* (June and July): 66.

Parker, B. 1986. Discovery of the expanding universe. *Sky and Telescope* 72 (September): 227-230.

Parker, B. 1986. The mysterious cosmic X-ray background. *Astronomy* 14 (February): 90-95.

Lesson 30

Barrow, John D., and Joseph Silk. 1980. The structure of the early universe. *Scientific American* (April): 118.

Bartusiak, M. 1987. The cosmic burp: The genesis of the inflationary universe hypothesis. *Mercury* 16 (March/April): 34-35.

Burns, J.O. 1986. Very large structures in the universe. *Scientific American* 255 (July): 38-47.

Darling, David. 1980. Universe: Time zero. *Astronomy* 8, no. 11 (November): 14.

Davies, P. 1985. New physics and the new big bang. *Sky and Telescope* 70 (November): 406-410.

Dicus, Duane, John Letaw, Doris Teplitz, and Vigdor Teplitz. 1983. Future of the universe. *Scientific American* (March): 90.

Kolb, E.W., and C. Quigg. 1986. Probing the structure of the universe from quarks to cosmology. *Physics Teacher* 24 (December): 528-538.

Lo Presto, J.C. 1987. The geometry of space and time. *Astronomy* 15 (September): 6-19.

Odenwald, Sten. 1987. To the big bang and beyond. *Astronomy* 15 (May): 90-95.

Rothman, T., and G. Ellis. 1987. Has cosmology become metaphysical? *Astronomy* 15 (February): 6-22.

Waldrop, M.M. 1985. A theory of the cosmos — With strings attached. *Science 85* 6 (May): 82.